Praise for *A Year on the Wing*

"I finished *A Year on the Wing* in something like a trance of pleasure. It was a struggle to leave. I think it has the makings of a classic [that] a great many people will enjoy and admire. It has something important and of course timely to say about the way we live (and don't live) on the earth. [T]he writing is frankly marvellous."
—Andrew Motion, Poet Laureate of the United Kingdom
from 1999 to 2009

"I think it's a masterpiece . . . I am full of admiration."
—Susannah Clapp, author of *With Chatwin: Portrait of a Writer*

"*A Year on the Wing* is a joy. Vivacious, engaged, full of knowledge lightly worn and generously shared—some lovely deft phrases and just the very words."
—Kathleen Jamie, author of *The Tree House*

"[T]he brilliant effort of the writing—that marvellous exercise of the imagination and language muscles making the sensual, visual thing come about again in words . . . and yet . . . never mawkish, no plangency in it, no pitiful precious sensitive little me—in fact, it's really funny, no boring protest about the modern world, just finding the place as it is."
—Tessa Hadley, author of *The Master Bedroom*

A Year on the Wing

Four Seasons in a Life with Birds

TIM DEE

FREE PRESS

New York London Toronto Sydney

Free Press
A Division of Simon & Schuster, Inc.
1230 Avenue of the Americas
New York, NY 10020

First Free Press hardcover edition October 2009

For information about special discounts for bulk purchases,
please contact Simon & Schuster Special Sales at
1-866-506-1949 or business@simonandschuster.com

The Simon & Schuster Speakers Bureau can bring authors to your live event.
For more information or to book an event contact the
Simon & Schuster Speakers Bureau at 1-866-248-3049
or visit our website at www.simonspeakers.com.

Manufactured in the United States of America

10   9   8   7   6   5   4   3   2   1

Library of Congress Cataloging-in-Publication Data
Dee, Tim
A year on the wing : four seasons in a life with birds / Tim Dee.
   p.   cm.
   Includes bibliographical references and index.
   1. Birds. 2. Bird watching—Anecdotes. 3. Birds—Psychological aspects. 4. Dee, Tim,
   1961-   I. Title.
   QL676.D267   2009
   598—dc22   2009010445

ISBN: 978-1-4165-5933-7

Contents

Introduction

Flying

. . . these were the first words
We spread to lure the birds that nested in our day . . .
LOUIS MACNEICE

The first bird I can remember watching flew through the garden of the house where I was born.

It is summer and I have just had my third birthday. I am pulling my red wooden train on its string; the train driver with his blue cap is swaying a little, because the grass beneath is bumpy. We are in the back part of the big garden of Acresfield, a Victorian house divided into apartments, on the outskirts of Liverpool. I am steering carefully because we are going along a thin strip between furrows of turned soil where the old man who lives in the apartment above us grows his vegetables. I must concentrate to make sure that the driver, who has a column of blue painted wood instead of legs, doesn't wobble too much, topple over, and roll from the train. Then I hear shouting and I look toward the noise: far across the wide lawn beyond the vegetable patch, the old man is leaning out an open window and waving his arms like a bear.

A year or so later when I meet Mr. McGregor in *The Tale of Peter Rabbit* I know him already. The man at the window seems too far away to be real, and I feel his voice must be loud and angry though it grows thin and falls toward the lawn. But he is shouting at me, and I don't like it. It is too much. I have to drop the string, abandon the driver and train, and flee, looking for my mother, heading for a greenhouse at the edge of the vegetables. I follow the path around a water barrel and keep going toward a wide black space of dark ahead of me across a gravel drive, the opening of the garden shed.

From behind me, over my head as I move toward the dark, flies a bird. It pulls up and into the dusty rectangle of the open doorway and disappears inside. It is showing me the way; I follow it.

In the sunshine, the space of the open door seems to be hung with a black curtain. I walk through it, and the air cools and the noises dim. The throat-catching smell of warm creosote comes. Everything is still. My eyes like the bandage of the dark.

Then, with a suddenness that makes me gasp, the swallow is there and then gone, diving down and out through the door space back into the bright. It calls once as it leaves, its buzzing twitter like an electrical spark. I look up through the murk and can see on a crossbeam a little mud pie with tiny sticks of straw poking from it. I forget my train and the shouting.

A nest.

That afternoon my father takes me back to the shed and lifts me on his shoulders so I can peer into the nest. Again, as we step into the dark, the swallow slips over us—so close I can feel the air rub against me. On my father's shoulders, I raise my arms toward the nest, slowing and softening my reach as I feel for the bumpy balls of mud and the prickly stems. There are no young birds or eggs yet. I can't see into the cup but let my fingers creep over its rim, feeling the smoothed lip and the feathers that line the tiny bowl. It is warm.

Some days later, I go back to the shed but find it empty and see on the hard ribbed concrete floor a square mess of baby swallow, a miniature hooked beak, downy balding feathers, raised but useless open wings, dead half-meat beneath the thin bat-skin.

I remember just these two scenes—one of calm and one of horror. I don't see the birds fledge any young; I have no concept of their departure. I cannot remember seeing them again. But I became a bird-watcher that summer. The swallows, their flight, their music, their stopped moments perched on wires or incubating their eggs, their nest, all this was somehow laid deep inside me, like iron in my blood, so that all swallows since the first one have joined that bird appearing above me and flying on ahead.

I have watched more than forty swallow springs and forty swallow autumns since those first swallows and their nest in the shed. All that time I have lived my life under birds and I cannot remember a single birdless day. Ever since then I have felt birds' rhythms answering mine, like a heartbeat or a stride, like a cuckoo's *cuckoo*. They come and go, they fly and land, they sing and call, they breed and die. Now, and then again. Locally they can get lost, go wrong, or be late, but birds fit the world; they are apt and at home. What they do and how they do it, the same over and over, gives their lives alongside ours an expression or a pattern in the air that can seem like art or ritual, as if they are deeper in the world than us, more joined to it, as we might dream it only. We have broken from nature, fallen from the earth, put ourselves beyond it, but nature, ever forgiving, comes towards us, makes repairs to the damage we have done. The swallow returns and builds a nest. Birds begin and end beyond us, out of reach and outside our thought, and we see them doing things apparently without feeling or thinking, but—and because of this— they make us think and feel.

I try to do both in these pages. This book is written from a life with birds but follows a single year of them, from one summer to the next; it begins with nests and eggs and chicks on the sea cliffs of the Shetland Islands in Scotland and it ends, a year later, with nests, eggs, and chicks in the holes of an oak wood on Exmoor in southwest England. I wrote it over such a year but I put into it many summers, many swallows.

I saw everything that happens in the chapters that follow—there are common birds as well as rarities and spectacles and remote terrain—but most of what is really my birdwatching isn't spelt out here, though it lies at the heart of this book. It is a birdwatching best explained by invoking the birds I am hardly aware of watching—the birds that are all around me, seen but not really noticed, there only in the corner of my eye, the pigeons and sparrows of the city. My day lists of birds usually stop with those species plus a gull or two. In some senses this book is about trying to make myself look at the pigeons, to go beyond simply noting them, to look at these birds (and so all birds) and see them as if for the first time but with eyes informed by years of seeing, to capture life looking at life.

As a teenager I was hard-core and obsessed, but I am not a professional bird-man now. I am not very sharp in the field. I misidentify birds and get bored and eat my packed lunch at half past ten in the morning. I do not want to fly, I am scared of heights, my palms sweat before every take-off. I am allergic to feathers too, and sleep under polyester rather than goose down. In other words, my birds are mostly the same as everyone else's, my basic birdwatching no different from that of most watchers of birds. But I fell in with birds early and I fell in deeply, and ever since then I have kept a kind of company that has taken me beyond birdwatching, back somehow to the pigeons and sparrows and that swallow at Acresfield, forward somehow to impossibly shared lives.

Here at the fountain's sliding foot,
Or at some fruit-tree's mossy root,
Casting the body's vest aside,
My soul into the boughs does glide:
There like a bird it sits, and sings,
Then whets, and combs its silver wings;
And, till prepared for longer flight,
Waves in its plumes the various light.

ANDREW MARVELL

June

Black Birds and White Nights

I will sing of the white birds
In the blue waters of heaven,
The clouds that are spray to its sea.
EZRA POUND

The seagull has no English
DAVID THOMSON

It is hard to sleep under a northern summer night. At midsummer in Shetland rain gallops on the roof. Outside, arctic terns and blackbirds keep up their noises through the darkless night, a troubling mix of scream and serenade, harangue and consolation.

I wake again and again to a pared-back half-light—like a prisoner's memory of day, my broken sleep crowded with a trapped dream of an impossible project. I had the task of launching into the sky all the seabirds from all the cliffs of Shetland: the shags and razorbills from the boulders at the sea edge, the guillemots and kittiwakes from the open faces of rock, the gannets from their shantytowns along ridge backs and fault lines, the fulmars from the tussocked *neuks* below the cliff lip, and the puffins from their burrows in the turf on the top. It was hopeless: no sooner had I flushed one wheel of puffins out over the edge than a gannet banked close to its nest and landed; kittiwakes dusted off in a cloud of bleached white, but guillemots flickered back, chocolate and cream; razorbills came, shags went.

Watching a seabird colony from a Shetland cliff top is not so different from my Shetland dream. Looking down and out to sea from the cliff edge at Hermaness (at the tip of Unst, the northernmost inhabited island of Shetland) or the Noup of Noss (the highest cliffs on Shetland's east coast; "noss" comes from the Norse word for "nose") you find yourself dizzyingly, exhilaratingly in the middle of a performance. Through every summer and for as long as the devastated seas can feed them, the wheels and flights of birds from their cliff places out over the sea and back again will continue. No one could get there before it started, and everyone will leave before it ends.

Without colliding, without even appearing to have to adjust their flying to avoid one another, thousands of birds are perpetually in flight. There are clouds of birds; there are wheels of birds; birds alone joining groups; other bird parties breaking up; birds landing; birds taking off; a line of shags heading out to sea; a line of guillemots rising to a cliff ledge; a carousel of puffins that collects and deposits participants on each, always counterclockwise rotation; gannets singly and in tiers, beating away from the cliff; fulmars spread-eagled on landing attempts and forced back into the air by irate neighbors. My eyes cannot stay on any one bird for more than a second or two; always another comes between me and the bird I am watching, always another flies behind it. I reach for words: it is a city, a metropolis, a theater, a cloud of gnats, Brownian motion, heaven, hell, space dust, bright stars in a dazzling universe—but nothing will adequately describe the colony. The only true account would be the thing itself. And it would have to include time, space, and sound. The colony is above me, below me, perched and in flight, far out, close to, huddled and singular, in burrows and beneath the waves. It begins with noise and smell and ends with light and quiet.

I hear, and gag on, the gannets at Hermaness before I can see them. Low clouds smear the rocks off the coast at the small, steep island of Muckle Flugga and the nearer stacks. The thousands of nesting birds on them are hidden, but the rain from the north brings in their fishy whiff in gobbets of stink, along with their homely rubbery conversation. Shafts of sunlight break up the rain for moments and I can see patches of open sea beyond the shrouded stacks. In every patch at every moment a gannet dives. I have watched them plunge into the sea thousands of times before. Again and again they do the same thing to catch their food, but each dive shines. Nature's repetitions are never boring. Every time it is like witnessing a fresh marvel in a new world: their visible decision-making, with its corrective twisting and corkscrewing, the rapid origami of themselves, and then their brilliant white match strikes, fizzing into the water (at 60 mph) to leave puffs of lit sea spray. It is hard not to blink and hold your breath as they go in. Against the hubbub of wind and rain and puffin coo and gannet vibrato it is impossible to hear them break through the sea, but the splash they make looks exactly like a sound and I find myself adding it to the ruckus.

From the cliff top I can see how the bubbling blue turbulence made by the brilliance of their dives traps the shape the gannet forms underwater. The diver pulls away and surfaces with or without a fish, but the cast of its plunge seethes like a tight net of ice just below the point where the gannet went in. It is rare for a bird to leave evidence, even fleetingly, of how it has moved through the world. It reminded me of how a collared dove left its dusted imprint on a window at home in Bristol when I was a boy. The bird had misjudged its flight from the bird feeder and briefly bumped into the glass. A ghost trace of the dove's raised wings was printed onto the pane by the oil and dust of its feathers. It looked religious and medical: an annunciation and the scan of a single flap. Unharmed, the dove continued to come to the bird feeder and for weeks, it and its flight print could be seen together, one through the other. The ghost lasted all summer, visible only when the sun was low, uncollectable, not even photographable, and I was sorry when the first heavy rain of autumn washed it away.

3

The rain at Hermaness has thinned and I can see further. The gannets pull up to the surface of the sea after their dives with a bucking torpedo energy. Every one I watch has to break out of a caul of water that their dive has thrown around them. They are hatching from the sea. I follow one bird as it dives, surfaces, and takes off, running along the sea to get airborne. I try to track it back to the stacks but lose it in the mind-wiping swirl of birds.

To see vast numbers of birds in silent movement is beautifully strange. In this crowd of so many moving things the eye leaps constantly from individuals to the whole event. These are not flocks but countless single birds caught in a storm of life. Individuals fly fast but the ensemble seems to be moving at dream speed. The rush of the single bird somehow slows in the crowd. I cannot focus for long on a puffin with its hurrying wings in the wheeling cloud; instead I see a tumbleweed of birds rolling out from their grassed lawns and terraces over the gulf of air, a slow-motion collective rotation. It drifts like the paper mobiles suspended over my infant children's beds that were blown above them through the night by their sleeping breath.

The noise of the cliff and the sea's moan down below is also silenced by these flights. On the ledges, the auks jabber and the gannets fret. Nearer to me, fulmars like miniature albatrosses *ack ack* as they return to their nests, and stout black-and-white puffins waddle along the cliff edge with matronly growls. But the flying wheels and lines of birds move with a filtered quiet. They are silent in flight. I can hear only the whicker of the nearer puffins' wings.

The gannet at sea and the gannet on its nest are hard to grasp as the same bird. In the air they move like great white flying shelves. They never appear to touch one another. But they must pair and breed and live within a beak-jab of others in a teeming midden glued to a dripping rock. They are big birds, with dagger bills and six feet of wings to fold away. They squabble and moan. Their nests are great bleached thatches built up over years, each merging into another, all held together with string and other irresistible rubbish the birds have collected at sea. They look like the nests of sea storks; guillemots, like house sparrows sheltering beneath storks' nests in Spain, breed underneath the gannets' eaves. The gannet is the scrap merchant of the ocean, its trash collector. A shag's nest, for all its putrid reek, is a far more organic affair. At the base of the cliff, the shags sit on their pyramid nests like kings of their castles, small cormorants, with their twisting black backs flashing a saurian green gloss. One gannet's nest has a blue plastic carrier bag carefully woven into its seaweed heart; another has a length of blue marine rope hanging down like a bell pull, and gannet neighbors occasionally lean across and tug it, as part of the continuing battles over space and furniture.

The sharp angles of the gannets' wings and beaks do half their territorial fighting. Their non-song does the rest. At sea, singing, talking, arguing, speaking of any kind has very little place in bird life. A gannet's face is silent. But on its nest it finds some half-aggressive, half-soothing noise in its throat, an elasticated gurgle, which works for love song and battle hymn.

Bonxies, or great skuas, which breed over the moor behind the cliffs at Hermaness and on Noss, have an even more comically bathetic voice. *"Bunksie"* means "thickset" in Norse and perfectly suggests their cunning—their business is the piratical harrying of other birds—and the heft of the meaty ballast they seem to carry. They are dog-food brown and chunky, the size of a fat buzzard, with mean and heavy beaks. In the air they are heavyweight bullies who behave like doormen to their own club, but on the ground they look dim and lumpen and there is something of a leftover dodo about them. Some call with a baby gargle *gu gu gu gu gu gu gu gu;* others manage a caveman *ug ug ug ug ug ug ug.*

I spot three skuas coming in off the sea at Hermaness. As they reach the beginnings of their breeding moor of bog, they raise their wings simultaneously into deep Vs and hold them hieratically for a few seconds. Then they come closer in a tough-guy balletic harmony and let off an *ug ug,* and I think of the sea world they have come from and the land world they have returned to.

We want birds to be wild and free, but we bring our ideas of them indoors——into our houses or into our minds——and we domesticate them there. Defining the skua calls as baby-talk and seeing their wings as heraldic really does not suffice. In my grandmother's house near Bristol, three ceramic drake mallards lived in permanent flight up the wall of the front room, their raised wings like those of the bonxies. But what could be less wild than those china ducks? To see a gannet stealing string from a neighbor, or a skua clucking at its babies? To know the seasonally grounded lives of these great maritime heroes, that they must land and coddle their eggs and sweet-talk their young, ought to stretch our mind. The skua isn't just the oceanic corsair, the gannet not just the fisher king. "Wild" does not describe nearly enough about the lives of these or any bird.

Puffins confound us, too. They come stooping out of their burrows like badgers and stroll at a post-prandial pace through the shivering pink flowering thrift of their front gardens. On the cliff top they seem at a kind of leisure. For a few sun-driven days of the year they are happy to walk with grass stems in their rainbow beaks. They dig tunnels, garden their lawns, take up flower arranging, and live like northern bowerbirds. They dance and croon and preen and kiss their lovers. They gurn at their rivals and joust with their multi-colored bills. It is hard not to laugh. Then the puffins fly out over the cliff edge, joining a rippling flex of themselves, and are taken utterly from the world we think we share with them. At every moment some slip from the turf and are shaken out over the sea, forming and yet always joining a spiral of puffins that rolls along the cliff. Sometimes a skua or a great black-backed gull drifts close to their burrows and a mass panic ensues; at other times, as some puffins land, new birds simply go. Their launches are briefly clumsy, as if a miniature badger is taking off. But once they are drawn up into their torqued spiral, their flight is superb. On any one wheel, all the available puffins are lifted and spun centrifugally in a continuous circuit three hundred feet across and a hundred feet deep. There are maybe four hundred birds all rotating in the same direction. What function these

spirals have is hard to know—the strategic befuddlement of predators perhaps, but it could just as well be a way of reducing collisions
on departure, or to practice for their whirring parrot flight, or—
dare I say it—simply fun.

I sit on the cliff top at Noss watching the endless, soothing helix
of puffins. By turns I resist and succumb to mapping the birds'
world onto mine. The puffin on the grass at its nest is where we live;
the puffin in the air is where we dream. We see the standing bird as
substantial; it has two feet, like us, on the ground. The flying bird,
by contrast, is transitory and adventuring. Flight thrills us because it
suggests the weightlessness we cannot achieve but also the freedom
of mind that we can. I am scribbling this in my notebook when a
splash of egg-yolk yellow puffin shit from a passing wheel-rider
puts a dribbling full-stop to my thoughts. It is still there now, like the
Shetland sun pressed into my page.

I rest my eyes and watch the sea and then, suddenly yet dreamily
slow, great fins come breaking through the water about half a mile
off the Noup. A family party of five killer whales is slowly crossing
the mouth of the great bay where the seabirds are crowded. Their
blows catch in a sparkle of sunshine. They bring up the wet of the
deep on their backs and show it to the sky and the auks and the gannets and me. Then they roll forward like dark planets bowled under
the sea. I try to follow them beneath the water, to feel their rhythm
so that I can predict when and where they will surface next. I get it
wrong every time.

I walk back around the cliff edge, chasing the whales, stopping
to look every time their fins come cartwheeling. The day seems
brighter still. The sea rises in the sunlight and smells and sings.
The whales go quietly through it, intent only on their swim. I feel
like a bystander behind a stage, glimpsing the backs of the main
act; or as if I am beneath a huge loom where the great skin of the
sea is being stretched over my head, the whales passing from one
side of it to the other; or like my five-year-old self, awakened in
the night, sitting in my pajamas, unseen at the top of the stairs

with my parents and their friends partying below me, laughing and drinking and dancing.

Watching the sea, I am waiting for the whales when something blurry and close crosses busily in front of me. I lower my binoculars. A tiny, earth-brown wren has come up from a narrow crevice splitting the turf at the cliff edge. It flies, calling over the green grass and the pink cliff thrift. Encouraged by the call, another appears, climbing into the air with less attack than the first. It cheeps and I realize that this is a first flight, a first appearance even, out from the crack in the cliff and the nest where it hatched. In a bumblebee whir it gets itself about five feet off the ground, then turns and flies head-on at me. I expect it to divert and wobble after the adult but it comes nearer and nearer until it is at my shoulder and the whirring and cheeping has stopped and it half clings, half perches to the strap of my backpack. It is four inches from my face. It hangs on for a few seconds, then flies off toward its parent, who answers its calls from a stone wall. For those seconds I have been the fledgling's stone wall, an indifferent wren at four inches and indifferent killer whales at a mile. The world was good and it didn't need me.

IT IS MIDSUMMER'S EVE. AFTER THIS DAY OF BRIGHT SHETLAND LIGHT, of puffins, gannets, wrens, and whales, I am going night fishing with storm petrels, waiting for that same light to fade enough so that the birds might come from the sea and bury themselves in the land.

Proper night lies off to the south. For weeks around the longest day, Shetland dims but never truly darkens. Though it has rained for almost all of my stay and the sun is still toweled by cloud, silvery light like developing fluid, laps down and the islands are floodlit. The time of day and the weather hardly matter. It is wet but it is light; it is night but it is light.

At eleven o'clock on that night-that-is-not, I wait with some other bird-watchers for a boat to Mousa. The small island to the east of the Shetland Mainland is still lit, golden and warm in the low light, with its ancient Pictish stone broch, our goal, looking like a tower of honey. It was built two thousand years ago to guard the sound and looks like a squat lighthouse or the cooling tower of an ancient power station. For the petrels to come, the broch must be dark molasses black.

A blackbird sings from the twisted fingers of a sycamore in a walled garden next to the sea. Tonight, with a handful of others through Shetland, these blackbird singers will be the last of their kind singing in Britain. Long after the rest of Europe sleeps, sounds of a suburban summer come from a few wizened northern trees. Elsewhere the blackbird's evening song is a fruited lullaby, but here there is no winding down to the dark, no last good-night kiss. The blackbird also sang like this at three in the afternoon, and it will gather its black cape of feathers and open its crocus beak into exquisite song all over again at two in the morning.

Farther up north it is even brighter. When I was nineteen, I spent midsummer's night with a girlfriend at Bodø on the Norwegian coast, seven degrees above Shetland and within the Arctic Circle. We had been arguing on the train north. There was next to no money left in our kitty and the loaf of bread we had bought in Copenhagen was finished. Why, Zoë asked, had I insisted we go so far? Just before midnight we climbed a hill, not speaking to one another. We sat on a boulder with a peeling elbow of orange lichen and looked out over the flaming sea. The sun swung low in the sky and hung there. Something in us calmed. To watch the last moments of a sunset and feel our world rolling away from the light is the most obvious evidence of the spinning earth. But here on the hill there was no setting sun, and so it seemed that time had slowed. The birds knew it too. Redwings were singing, my winter thrushes, the birds that I had heard calling *seep* to one another flying over me in the dark as I walked home from Zoë's house through the streets of Bristol in February. Before Bodø I had never heard them say more than *seep*. Now, from the dwarf birches just below the crest of the hill two males sang an evening song, unwinding, fluting, and sad. As we watched the sun idle for thirty minutes, they fell silent. Then, knowing as we did that it was time to stand up and start the next day, they sang a dawn song, tightening, perky, and assertive. The sun followed and pulled up through the peach sky it had made. The redwings' night was done in half an hour of bright light. Zoë and I walked down the hill happy and restored.

Around midnight on Mousa in June, the sun has gone below the horizon and its light is turbaned enough for the storm petrels to come to their nests. We cross the sound to the island in an open boat. The ferryman in his sailor's cap leads us from the jetty along the shore, though the path is clear to see. In the stretching twilight we arrive at the broch and stand peering up at its darkening bulk, not certain what to expect. I stoop under the low lintel of the sea-facing entrance and through the six-foot-thick walls. Inside the roof has gone and it is like standing on the hearth of an old chimney rising to the night sky. Between the outer and inner dry-stone walls, a stone stairway spirals to the top. I begin to feel my way up and have climbed ten stairs when I hear the first of the petrels singing. I stop. In an instant, night has fallen.

Storm petrel song is the strangest of music. The petrels breed in tiny chinks between the stones of the broch and from its curving wall comes the sort of squeaky skirl that the stones themselves would make if they could rub together. A second bird makes a start-up burp a few feet away, and then another, though it is hard to tell where the sounds are coming from or to imagine how a bird could be making them. I stand in the dark on the spiral stone stairs, listening to rock music inside a giant's ear.

At night, birds' night songs sound ancient, weird, and dark: the reels of grasshopper warblers, the grunts of woodcocks, owl yelps, corncrake rasps, nightjar churrs, even the tearing-apart music of nightingales. All these noises are chthonic: they come up from a very old earth. The storm petrels' song is even weirder: a fossil and furry pulse from the back of a tiny throat hidden in a crack in a rock. It is an inward and sibylline sound of swazzle notes and speaking stones. It giggles and rewinds. It clucks and purrs. It sounds like a simultaneous possession and exorcism. It is a noise reserved for the night and the deeper dark of the petrels' nesting holes. It would be unimaginable in daylight and inaudible over even the quietest of seas. It defines the deep contrariness of the midsummer storm petrel.

Shrunken birds, storm petrels are little larger than house martins and superficially similar: their flight is buoyant and they have black backs and a white rump. In the vast space of the sea, they are so small that they are very hard to find. Only after hours of looking, when crossing the Bay of Biscay once on a huge ferry, was I able to spot some petrels, like particles of fluff, pulling their tiny bodies through the troughs of the sea far below me. But at sea, the neighbors of fin whales and sunfish, the petrels live far more happily than on land. Like midget puppet-masters, they seem to hold themselves at a constant height above the water, as if pulling its waves into tides. They are so slight that the sea appears unable to hurt them. They fleetingly dip down with delicate attentiveness to patter their black webbed feet on the sea, tiptoeing over the swell, giving it the gentlest of massages, walking on the water. And from this pattering miracle, and St. Peter who did it too, they take their name.

Away from the sea they are disastrous. Seabirds blown onshore or inland are called 'wrecks' by bird-watchers. And so they are. In the spring of 1850 a boy caught a moribund black-capped petrel that he saw "flapping for some time from one furze-bush to another" at South Acre, near Swaffham in Norfolk; "exhausted as it was, it had strength enough remaining to bite violently the hand of its captor, who thereupon killed it." The bird was skinned and stuffed and until 1984, when a corpse washed up in Yorkshire, was the sole British record of this oceanic petrel that breeds on Hispaniola but prefers the pelagic life, far out in the Atlantic. In December 1999 forty storm petrels (including birds ringed on Fair Isle in Shetland and Skomer in Wales) caught up in vile weather were forcibly displaced from the Atlantic to the Jura Mountains in Switzerland. Try to imagine their journey. At the end of

October 1952 more than two thousand Leach's storm petrels were funneled into Bridgwater Bay in the Bristol Channel and wrecked there, half of them dead. I saw my first storm petrel in September 1977 in the same place, on the miserable brown shore of Steart, clouded in the stench of the chicken sheds that cluster at the beach. A broken and blasted bird, its compass shot by the great wrath of a western gale, it tottered on the tidal mudflats, looking like a cripple whose sticks had been snatched away. The bird that can walk on the water is lost on land. All petrels, the world over, are like this; the sea is their home.

The broch at Mousa is only fifty feet from the shore. We stand in a ring and wait. Eventually black noise from the dark tower coaxes black things from the dark sea. A first bird appears a few feet above my head and is gone in a pixie bat flap. It flickers and jerks around the tower, as if it were being projected onto it in a damaged home movie. Another bird comes and then another. They seem to rush at the broch, reeled toward it on invisible threads.

The petrels seek a specific crack; some cracks sing and others are silent. The arriving birds are quiet but for their flutter. Can the incomers identify the squashed yodel that they need? A mate's song among all the hiccups and space noise? Do both sexes call like this? Will the singer leave when the wanderer returns?

Petrels move over the land as though it is repellent to them. Their flight suggests that this necessary short journey from the sea to the shore is forever disorienting. It is similar to how we walk up church aisles—as if we have stones in our shoes. Some petrels find the hole they are looking for right away and scramble in; others begin to spool around the tower, some bashing into the wall and crash-landing or selecting the wrong hole and falling backward down the side of the broch. Two birds collide at a single hole, squeak and scuffle, and one of them tumbles to the ground at my feet. As I reach down, it spreads its wings into the springy turf and manages to launch itself back into the swirl. Another brushes so close to my left shoulder that I can feel its wing beat, a tiny push of air in my ear. I begin to smell them. Squid and salt.

Of the birds I can see, most are being pulled to the broch and its sooty night-light but others seem to be circling it more widely. Have these birds swapped egg duties and come out from the tower? Three days is the average storm petrel incubation shift, three days in a damp, black crevice. Then, after a rhapsody of squeaks, they tumble out into the dark night air, toward the sea and quiet. A midnight flit.

I sit down on a boulder at the edge of the beach. A petrel churrs from beneath me. Much of the island is noisy with them; there are six thousand pairs here. The beach snickers and the field walls cough dry, scratchy tunes. We walk back to the boat along a purring shore. About one in the morning the dark begins to ease, and tomorrow yawns awake before we have gone to bed. At five past two, back on the mainland, the blackbirds are singing again. Another black bird, a different midsummer night's dream.

Driving in rain the next day I stop at the oil terminal at Sullom Voe. Various hard-core birding friends and acquaintances traveled to Shetland to work on its construction in the late 1970s. The money for laboring was good, there was little to spend it on, and there were wonderful birds—especially for southern birdwatchers—to see. Snowy owls bred on Fetlar; the lone black-browed albatross was still trying its luck with the gannets at Hermaness. Both have gone now, having failed to make their way, and these days, although the terminal still serves as the landfall point for much of the ebbing North Sea oil, the place seems near ruined. It has the feel of Eastern Europe and the militarized Soviet world, when rust took charge in the 1980s.

After Mousa broch and its singing stones, I am struck by the weariness of the terminal's buildings. The crumbling sign at the abandoned gatehouse reads SU L M VOE TER IN L and I couldn't get "petrol" and "petrel" out of my head: black stuff coming out of the sea, one exhausted within a human generation, another living on in a two-thousand-year-old tower block.

The next day, leaving Shetland, sitting on the runway at Sumburgh Airport, my mind roves toward these thoughts and away from them, accompanying my sweating palms (my fear of flying) and my squint at some gulls and terns above a pier. We take off south over a churning sea, past Fair Isle and Orkney. I read some of David Thomson's extraordinary book *The People of the Sea*, with its seal myths and stories of meetings of seals and men and how these survived in Britain and Ireland into the twentieth century. Midsummer was one of the times when seals turned into men and women. Flying over the seals and the whales and the seabirds, I feel gloomy about our separation.

I fall into conversation with my neighbor, the friendly manager of a new supermarket in Lerwick. He tells me of his banana problem. We have become a nation hooked on bananas, but bananas on northern islands are not easy. Their usual life span on a supermarket shelf is only two days. They arrive refrigerated from the tropics in a state of arrested development. Opening the box that they have traveled in releases a ripening agent but from that point the clock ticks fast on their salability. A banana reaching Shetland would have to be sold the day it arrives and this is too risky for the supermarkets given the bucking seas that surround the islands and hamper shipping. The solution is to give Shetland the only bananas in Britain that will ripen without assistance.

Once, on another Scottish island at a low tide, I had watched a seal curl its head and feet flippers into a gray smile that looked like a ripe banana. Hispaniola—Haiti and the Dominican Republic—grow some of the bananas that we eat in Britain, and somehow in 1850 an unrefrigerated black-capped petrel had gotten from that island's tropical slopes to a gorse bush in Norfolk. Meanwhile in transit at sea that day, beneath me and the supermarket manager, were all the storm petrels that had been released from their egg-ripening duties on Mousa; somehow these birds and those that had swapped places with them at midnight, and their as-yet-unhatched chicks, would all be flying to the south Atlantic in a few weeks on their six-inch wings.

July

Neither Sea nor Land

The full moon glided on behind a black cloud. And what then? And who cared?

S. T. COLERIDGE

Rusham Road, 6–7.15 a.m. one thrush hammering away at one triple cry, message or whatever anyone else likes.

EDWARD THOMAS

Every day of bird-watching is different. I have been to Wicken Fen (a few miles north of Cambridge) many times before, but this evening in early July it is more landlocked than ever before. A warm wind blows as if from an open oven and pushes baked dust, pollen, insects, and seeds into the corners of my eyes, my ears, and my lips.

Toward dusk, I walk from the fen into a wood of young willows. Cuckoo-spit—the frothy protection that insect froghoppers secrete around themselves—drips onto me from lather peaks like beaten egg-white that smears the joints of the twigs and branches. The spit is heavier than rain, gummier and more viscous. The droplets make a soft *silp* sound as they land. I can hear the whole copse thickening with spit.

The floor of the wood is carpeted with the drifting down of willow seeds and their fine-spun woolen nests. Spiders' webs that have trapped the down look like overdone Halloween versions of themselves. The wool and the spit. It is like putting on a wet pullover on a hot day.

A cuckoo at the edge of the trees and another down a ride—a pathway through the wood—start cuckooing. They turn their heads and throw their voices. The fen deepens still further. The calls, like cupped heartbeats, seem to arrive from far away in time and space. Each cuckoo apparently holds a mirror that sends back its rival's call, blanked and thwarted. They sounded dusty and muffled, coming as if being blown, as Gerard Manley Hopkins described them, through 'a big humming ewer'.

It makes for a claustrophobic broth—the cuckoos, cuckoo-spit, and summer snow—and it drives me back out onto the fen where it is easier to breathe. The heat fades and the sky opens upward. The day lives on in the clear upper air and light. Swifts and swallows clean it by dabbing at its insects, singing as they fly. The swallows' garrulous static ripples and phases as if the birds were mediums channeling the sounds of the sky. Higher up, the swifts scream their mastery of flight into ecstatic shouts.

I walk through the fen, waiting for it to get darker. The day is reluctant to finish. Two common terns make last flights above the reedy mere, white as ice cubes against the green. In a hedge along a dike, bullfinches pipe their embarrassed music, their soft calls of bloodied regret escaping over their blood-red breasts. Sleepy cormorants fly into the heronry in a tall alder beyond the mere. Their ragged wings, black with a used sheen like an old man's suit, made gashes in the spreading shambolic tree.

The light slips higher and the darkness rises. A minute before, the low sun had picked out individual stems of the reeds; now, they mass into bigger and bigger plots of dark. A little egret floats to its roost, through the swallow's air and toward the setting sun, looking like a flying lantern that has trapped light in its brilliant phosphor-white plumage and spilled it into the darkening sky as it flies, or like the golden pin of an airplane high overhead still bright long after such colors have drained from the earth. As the egret sails away, a thirty-year-old thrill and relief at having seen my first in Britain comes back to me. As a teenage rarity hunter, I took a complicated journey by bike, train, and bus, and several miles on foot, before a distant hunched white sheet on the edge of a marsh near Portsmouth came into view. The egret's white was then a color not known in Britain. Now it is commonplace.

Below the egret at Wicken is its antithesis, a hidden bird making a noise like a small mechanical pump. The singing grasshopper warbler sounds as if it is drawing more and more dark up to the surface of the fen. Others have heard a sewing machine, a spinning wheel, or a fisherman's winding line in its insect stridulation, but for me the grasshopper warbler's song is the silvery purring of the earth itself, its dark warmth expressed as a reeling drone at the end of the day.

The crepuscular and hidden habits of grasshopper warblers tie them to a few inches of life just above the wet soil of the world. Inconspicuously drab and streaked olive brown and buff, they are like a color chart of the time of day they sing in, the thinning tones and lost pigments of dusk and night, and of the habitat they live in, their winters buried in the grasslands of Senegal, their summers hiding in the reeds of Wicken, and their night flights of migration between. You must catch a grasshopper warbler while you can. I might only have a handful of summers left when I can still hear them. They are lost to my father; their song is now beyond his hearing range. If you haven't heard one before you are fifty you may never. Seeing them is no easier. They are in Britain just for a few weeks each summer. In their silent daytimes at the bottom of a reed bed or scrubby ditch, they are invisible and often start singing only in the last minutes of light. And when they sing they are usually hidden.

At Wicken two birds sing, the near male in purring battle with a neighboring one, both hidden, to define and contest the invisible boundary of their territories and to attract an invisible mate. Here, in this bird that seems to be spooling time with its dancing reel played out and wound in again, is proof of Keats's line "The poetry of earth is never dead." The grasshopper warbler's song comes up out of the earth just as you think the day is quietening down. It has no prelude or warm-up: it starts midperformance, as if switched on by the dark, as if it has been singing like this beneath the earth until it was dark enough to come to the surface. As if, though we hardly hear it, this singing never stops. The near bird falls silent but the answering male continues across the fen. I wait and stare through my binoculars at the murk of reeds and bushes, but I can't see either. The singing continues. The night shift has begun.

I retrace my steps toward the dripping willows. Night has all but fallen and there is next to no color left on the fen. The dark blue of the sky above the copse has deepened to a velvet stage curtain, and across it is flying a woodcock. With the kerfuffle of a Dickensian clerk, hunched and wheezing and crabbed, he looks like he is carrying a heavy ledger up a hill. It is hard to think that he is doing a sex dance. Against the sky I can't make out any colors on him but see only a cut-out shadow puppet with rounded wings, a round thickset head, and a long bill angled to the ground. The puppet has a strange jumpy progress and rocks from side to side, a bird maneuvered on sticks or hoisted aloft and held up there by some force, an evening gas or a will-o'-the-wisp that has risen from beneath him out of the same wet marsh soil that he lives in and that has pushed him skyward like an accidental balloon. But go up he must, and above the ride cut through the willows, the woodcock rodes.

As with much else about it, the woodcock has this word—"roding"—to itself. Other birds display during the breeding season at dawn or dusk; the woodcock, alone, rodes. The word seems to come from the old practice of trapping woodcocks by setting a net within a wood along "cockeroades."

The bird above me makes little *hisp* noises, almost under its breath, like the drip of the cuckoo-spit, and then an even quieter, rather filthy, wormy grunt that transforms it into a flying pig and ties it to the soil and the leaf mold where it will live for the rest of its life. These two noises, a snore and a sneeze, are the bird's roding song (they have been transcribed as *"quorr quorr-quoroPI-ETZ . . . quorr quorr-quoroPIETZ"*). The piggish snuffle might also be the sound they make when their long straight snipelike bills probe feelingly into the black mud, like a blind man's stick, reaching after worms, seeing in the dark. John Skelton in his poem "Philip Sparrow" lists "the woodcock with her long nose," and *The Handbook of the Birds of the World* describes the bill as "rhynchokinetic." The flexible tip of the woodcock's upper mandible has been said to resemble more closely "the writhings of a worm than a beak." It uses its worm beak to feast on a diet of worms, eating almost its own body weight of them in the course of a single day.

I have never seen a woodcock bend its beak, or eat a worm, or even on the ground at all; yet the earth defines them absolutely. They feed in it, breed on it, and are colored like it. Flying, they seem broken from it. As if the brightness of day is anathema to them, they only appear when the sky grows earth-dark. Yet their odd limping display is a dance to be seen, and for all the woodcock's weirdness and my sense of them having been yanked from their preferred place, their roding flight is bewitching to watch. It manages to be alien to the bird and yet entirely consistent with it. Roding woodcocks can also slow time. As I watch three birds going overhead, or one bird on three rodes of its territory, the fall of light is stalled by the slow rocking flight. Night waits for them as if, in some ancient ceremony, they are the ushers of dusk.

Eating a woodcock, as I did once, was like eating the earth. No wine has ever released its terroir to me as that bird did. It tasted like a prune, sweet and sour at once, a mixture of loam and chalk. The bird's dark purple flesh crumbled on my plate like a dried worm cast and the worms that made its meat. I had held the bird in my hand before it was plucked and cooked among the white surfaces and steel utensils of a smart London kitchen. Its cryptic moth-wing colors gave it the look of a worn fireside rug. To hold its book of browns, the wings falling open on either side of its body, was to sense the humus of dead leaves mulched into a bird over thousands of years—the woodcock as a surviving fragment of an old earth, from a time when leaves became birds, branches grew wings, and the dark moved. It shares this quality with only two other birds, I think: the wryneck, an aberrant woodpecker, and another night bird, the nightjar.

Unlike these two, and because it is tasty, the woodcock has been much hunted and much written about. As I write I am sipping a burgundy with a woodcock on its label, an earthy pinot noir from the Cave des Vignerons de Buxy in the Côte Chalonnaise. I browse the entry for woodcock, or *bécasse,* in the *Larousse Gastronomique:* "Its plumage is the color of dead leaves; it has long been regarded as a delicacy." I try to imagine tasting casserole of woodcock, or casserole of woodcock à la crème, or cold timbale of woodcock, or cold woodcock à la Diane ("Pound the intestines with a knob of foie gras, a knob of butter, nutmeg, and brandy"), or hot woodcock pâté à la Périgourdine, or roast woodcock on toast, or sautéed woodcock in Armagnac, or truffled roast woodcock, or woodcock casserole à la Périgourdine . . .

We roasted the woodcock that I ate like a chicken. Plucking it sent a leaf-drift of earth into a steel sink. We bent its head beneath it and stuck its ivory bill through its thighs so it trussed itself into an avian pietà. I ate half its breast and a leg and picked a ball of lead from my teeth.

I don't' know where my woodcock came from but the biographies of others, even from long ago, are readily assembled. We can put together the end of a bird's life, in human words if not woodcock terms. In September 1465 a banquet for a new archbishop of York called for 400 woodcocks (along with 2,000 geese, 204 cranes, 4,000 mallard and teal, and 12 porpoises and seals). Gilbert White's journal entry for October 1, 1777, reads: "Bright stars. This day, Mr.

Richardson of Bramshot [in East Hamsphire] shot a woodcock: it was large & plump & a female: it lay in a Moorish piece of ground. The bird was sent to London, where as the porter carried it along the streets he was offered a guinea for it." Alec Douglas-Home, writing in 1979, has an amazing tale to tell worthy of a saloon bar or an after-dinner speech, or a second bottle from the Cave des Vignerons de Buxy: "We were shooting above a railway cutting when I shot a high woodcock which fell into the tender of the Flying Scotsman. I just had time to signal to the driver as he flashed past, and he left the bird for me with his compliments with the station master at Berwick-on-Tweed." Elsewhere freaks as well as flukes and feasts are recorded: "In the year 1833, a Woodcock with white feathers in the wings was observed in a cover on the manor of Monkleigh, near Torrington, in the county of Devon. The same bird, or one of exactly the same plumage, reappeared in the same place during the four succeeding seasons, in which period it was so repeatedly shot at by different persons without effect, that it at last acquired among the country people the name of 'the witch.'" In the year 1837, however, it was killed on the property of a minister (the bizarrely named Reverend Pine Coffin), who had the specimen preserved. In eighteenth-century Cumberland, William Wordsworth hunted woodcocks as a child. The first memory he recounts in the celebrated passage in "The Prelude" that begins "Fair seed-time had my soul, and I grew up / Foster'd alike by beauty and by fear . . ." is about trapping woodcocks (and later stealing other trappers' birds), and he describes getting close to them with an uncanny mimesis that replicates, at least in my mind, the woodcocks' stop-start walk—a walk I have never seen.

> . . . 'twas my joy
> To wander half the night among the Cliffs
> And the smooth Hollows, where the woodcocks ran
> Along the open turf. In thought and wish
> That time, my shoulder all with springes hung,

I was a fell destroyer. On the heights
Scudding away from snare to snare, I plied
My anxious visitation, hurrying on,
Still hurrying, hurrying onward; moon and stars
Were shining o'er my head; I was alone
And seem'd to be a trouble to the peace
That was among them.

Roding is not the beginning and end of woodcock flights. The summer's reluctant flier is actually a seasoned mover. Softness of soil is vital for woodcocks, and the freezing woodland floors of northern Europe drive them into the autumn air toward the milder south and west. Most British woodcocks are sedentary, but it is a migratory bird for much of its range, and many thousands join ours each winter. (Those in North America winter in the Southern states and Caribbean.) Cornwall and the west of Ireland, where frost is rare, receive high numbers of northern birds. One February dusk I saw one drop hurriedly from the blasting Atlantic sky into a tiny garden of three or four stunted trees on Bryher on the Scilly Isles—about as far from the unending forests of the taiga as I could imagine. Gilbert White reports a great October fall of the birds on the same islands: "some woodcocks settled in the street of St Mary's and ran into the houses and out-houses," and "a Gent . . . shot and conveyed home, twenty-six couple, besides three couple which he wounded, but did not give himself the trouble to retrieve."

Woodcocks fly fast and direct on migration; so fast "that a pane of plate-glass of an inch thick has been smashed by the contact, and one was actually impaled on the weathercock of one of the churches in Ipswich." Goldcrests—the smallest European bird (similar to golden-crowned kinglets in North America)—which share the northern forests with woodcocks and likewise need to get out in winter, were once thought too tiny to manage flight for any distance and certainly not able to survive a crossing of the North Sea. They were imagined as woodcock pilots, riding on the back of a brown-

and-ocher flying carpet. Woodcocks themselves have long been thought to carry their own nestlings between their thighs, away from danger or to water. Many bird-watchers nowadays don't believe this, and today's textbooks turn up their rhynchokinetic noses at the idea. *The Birds of the Western Palearctic* sounds a typically cautious note, recording a flushed female woodcock "calling and apparently struggling to prop young between feet with depressed tail." Earlier authorities were more generous. One is a staunch believer in the woodcock as porter: "I have several times seen them flopping away with a chick gripped between the thighs, and once with chicks on the mother's back." Several other accounts report, "She rose with him in her feet, her long legs dangling and swinging with her little burden like a parachute" and "the young bird was clasped between the thighs and pressed close up to the body of the parent"; "the woodcock supported her young not only with her feet, but also with her bill pressed over the chick against her breast." I imagine a woodcock Madonna and Child.

As a young bird-watcher, my note-keeping swung from the baldest of logs and the most austere of lists to winding prose paragraphs that aped accounts I had read in bird reports and books, describing weather conditions, the exact route I took along footpaths to local ponds, and precise counts of common resident birds. My first woodcock record is spare. The entire notebook entry reads: "7th July 1974 Thursley Common, hot, woodcock 1." There is not much to go on, yet I find I can rebuild that bird from those eight words. I was thirteen. It was a hot summer's night. My father and I had gone to the same place where a few years before on a bird club outing we had been shown a nightjar. This time we went home before the nightjars got going. As we walked down a dusty track through gorse and toward a birch wood, a woodcock materialized above the trees, silhouetted against the dusk. It hurried along the skyline for a few seconds before turning away and hiding itself beyond the wood. It was exciting but disconcerting. I was sure of our identification, pleased with how like a woodcock a woodcock was, but it offered so little of itself that I felt I had been simultaneously shown the bird and excluded from it. Later I came to recognize this sensation as important and continuing. It is the defining condition of most bird-watching.

Two years later I saw another woodcock in Norfolk in eastern England. I watched one fly in off the sea over the Wash, haring down through salted gray October skies. It was half a millennium away from the four hundred that the archbishop had sacrificed. Seeing this woodland snipe, a piece of airborne leaf-mold, flying from the sea toward the security of some wet muddy spot made me want to understand its past. I began to reverse its story from the point where we had met, sending it back across the unforgiving North Sea and over the mountains of western Norway, then high over all of Scandinavia to the northern forests of Karelia on the Finno-Russian border. There, in my mind, I made it a nest in the soft summer earth of a forest floor, a woodcock-breast-shaped scrape lined with ferns. I gave it eggs and made it sit on them. Then the bird came toward me again, the eggs became chicks and fledged, and the adult woodcock left them. It lifted up to the autumn sky, crossed the Iron Curtain at night, and flew over the Baltic and on to our twenty seconds together, five hundred feet apart, under the icy drizzle of north Norfolk.

THE WASH IS THE GREATEST OF THE NORTH SEA'S JAW BITES THAT open the body of England. In late July, I am in a place as placeless as the open ocean, edgeless, hard to enter, harder still to hold, and full of hidden things and distant birds. It has the color, consistency, and mobility of slip, the warm brown muddied water used to keep things wet and moving on potters' wheels.

The Wash's huge square yawn looks big enough to drain everything that flows into it, but it gargles instead, holding all the silt of the East Midlands, the Fens, and much of East Anglia. To speak of the mouths of the rivers (Witham or Haven, Welland, Nene, Great Ouse) that drain into the Wash seems absurd; *it* is the mouth. Its thirst cannot be slaked. All the rivers run into the sea; yet the sea is not full. Sometimes the Wash will gulp down what pours in; but just as often it will spit out what it has half swallowed. On one tide the fens nourish the Wash, on the next they might be force-fed by it.

Following Peter, the fisherman who has invited me to go out with him, and walking north, I wade through a mud bath. As I slosh up to my thighs in the retreating tide, the mud seems to have come up from my childhood. The brown cloud I kick eventually settles like a dreary summer's day on a wet British beach of not-quite-sand. The water is so murky that I can't see my feet. Like a drunken dowser I stumble and lurch, feeling my way down channels of sucking mud, bumping between ridges of slippery mud. I want water, but there is just mud. Where does the shore become the sea?

A stick shows above the wave tops, marking the land end of the fishing net I am headed toward. It seems impossibly distant. I can't believe we will be able to walk that far into the water to reach it. But we do, and entangled in the first net ahead of me, a fish flails at the surface as tidewater draws away from beneath it. The net has worked; there will be a catch.

It is raining and we have to wait for the tide. Peter chose a neap, but the north wind blowing straight down the open mouth of the Wash has pushed water higher up the salt marsh and is holding it there longer than usual. Held in a great circle of wetness, Peter talks about his Wash: his family on its shore and how he is drawn to its water again and again, the wildfowl he shoots, the fish he catches, the samphire he gathers. "Samphire"—the word comes from the French, *"l'herbe de Saint Pierre."* Peter and petrels and St. Peter and water-walking and samphire swim though my mind.

We shelter in the lee of a man-made mound like a great burial round-barrow or earthwork that has pushed the marsh farther out into the sea. This is a burial chamber of human ambition, the moraine of an abandoned experiment to create a freshwater reservoir at the edge of the Wash, just one of a succession of interventions that have tried to keep things still in this extraordinarily mobile place.

All is temporary on the Wash. This is a wilderness made of mud suspended in water, perhaps the last wilderness in England and surely the biggest. Its tides live for twelve hours; it is born and dies twice a day. All life and all energy is in the great estuary's tidal movement, from mud disappearing to mud reappearing, in the passage from the edge of the marsh to the sea, from freshwater to salt, from sea purslane to samphire, from green marsh to intertidal mud, from muddied brown sea to more-salty gray sea, from sea to sky to cloud to rain. All these zones are porous; each leaks into the other.

The cycle of inundation and retreat is endless. The throat of the great mouth can be either the blue-gray open marine sky of the northern horizon or the black soil of the fens to the south. Conducting all the business between the one and the other is the mud, a huge tongue lolling in the mouth. Imagine a whale's tongue gray-brown and wet forever, fifteen miles long and fifteen miles wide—that is the mud of the Wash.

The conditioning color here is brown. Everything runs and nothing is fast, but brown dominates from the end of the fields to far out to sea. It is a brown like a jam jar of water after a child's busy painting session, a brown that contains all the colors: the slops of the fens draining green and yellow crops; red poppies; ruddy dust carried from brick pits fifty miles away via a life in the breeding breast plumage of black-tailed godwits; black soil laid into the heart of heads of green celery; the polar white of Bewick's swans, brought to the Ouse from arctic Russia; pink-footed geese's pink feet ripening like tundra berries beneath them; the shelduck's umber; the dunlin's dun; even the hated, escaped, and murderous mink's black-brown stole. All these raw colors are washed down as silt, salted by the sea, then join the mud. Add any more you like; everything becomes brown.

As on any huge estuary around the world, you have only to be on the Wash for a day to realize the futility of human plans for any permanence in the place. What you take to be a fixed horizon is picked up by the wind and disappears. Water vapor condenses just above the fields of wheat and yellow rapeseed at the shore and deliquesces their edges into cloud. A whitethroat flies out over the salt marsh from its grassed nesting bank on the most recent seawall, singing its dry ratchet song over the slippery green ooze; a redshank agitated by a marsh harrier towers inland over emerald wheat fields calling its bleak mud-flat alarm. No edge is defined. This is a place that is no place, that is never what it was or will be. Everything gives and takes, slides and slips.

Peter and I walk out across the mud and into the water, but it would be just as accurate to say "into the mud and across the water." I look around; I am standing in a sea. The thin line of the marsh and the open mud we started from now seems so distant and so slight on the vast horizon that I laugh. Two stick figures seeking a third have walked determinedly into the water. We must look insane from the shore, except there is no one there. There is no one as far as my eyes can see. We are alone, standing in the lumpy brown water as the tide battles around our legs, the wind pushing water south toward the land, the invisible moon sucking it away north for its duties elsewhere on the planet.

In the nets are a bass and six gray mullet. We let go the two small flounders, or butts, and they snake back into the water like miniature waves, mud-brown above, sky-gray beneath. The mullet have tangled themselves; Peter unhooks the nylon filament from their bleeding gills and popping eyes of fish panic and eases the fish back through the net. Most are still alive when we reach them, and being released from the nets gives them a little more twitch. Peter takes the wooden fish priest, or club, strung on his waist and hits each mullet very hard three or four times on its head. It is bone crunching, a loud and finite noise amidst the never-ending ripple and blow of wind and water. As he hits them, Peter talks to the fish in a soft voice. "I know, I know," he says. After each of the whacks, their bodies clench and flex against his arm before they return to their fish shape. When they die Peter lays them gently in his shoulder bag. By the time we have emptied both nets the tide is at our ankles.

Gutting is next. Peter turns the mullet belly-up, takes his knife, and pushes it through the fish, just below the gills. He cuts bones and pulls the blade up toward the skin so the fish snaps open one-third of the way along its body; now he can slide his knife down the fish's belly, opening it in a great split. The gut in all its warmth and reek bubbles in the sluice of the split. A green vegetable slush creams out, the mullet's final meal. The viscera spill into the sea. Four herring gulls detect the offal one hundred and fifty feet up in the wind. They call and jilt in the sky but won't come down while we are there; perhaps they have noted it for later. Peter bends to the sea and washes the gutted fish, his knife, and his hands. The clouded water clouds some more.

The sky is fine muslin; the gauzy drift of rain from the north lifts and takes the horizon up with it. I look out toward the Lincolnshire shore and can see it twice, layered on itself. A mirage has raised a line of trees, floating them off the ground into the lower sky, as if to emphasize the impossibility of ever arriving there or reaching any-where that would be a real edge to this place. Farther north, the church tower of Boston Stump hangs detached in the sky, a blurred color field of brown in gray.

I have seen this effect before, though not when standing in the sea. The *puszta* with its oceanic grasses and shimmering reeds at the center of the great Hungarian plain is an inland dry Wash. It is so flat and has such a wide horizon that you can see, it is said, the cur-vature of the earth. I think I saw that, but in the same place I also saw giant stretched cows sailing through the air with horns prod-ding the heavens, a magnified image of what lay beyond the hori-zon, thrown into the sky and projected on its vast screen, a *fata morgana,* or in Hungarian a *délibáb.*

I have often wondered whether flying birds might see things like this, being able to look beyond our horizon, and if so how discon-

certing a giant floating cow might be to them. That way Don Qui-
xote's visions lie, I know, but standing in the Wash looking toward
Lincolnshire I am also reminded of the poet John Clare, saturated
with looking at what was close at hand in front of him but wandering near the Wash, to the borders of his knowledge, getting to the
"edge of the orison" and thinking he might look over the lip of the
world. On the Wash, it seems you can.

Peter is alone out here but not lonely. His wife worries about him
when he is on the Wash, but he likes it. Keeping his eye on his work
in front of him, untangling the fish and dispatching them, he doesn't
notice his isolation. As he gets on with scaling the mullet, a solitary
gray plover flies over us. It calls once, a lone wader flying across a
wet summer sky. Most of the gray plovers that winter on the Wash
have finished breeding now on the lowland tundra of high Arctic
Russia, but this one hasn't gone. It won't go north this year and
won't breed.

Flying here alone across the pallid Wash, its plumage of courtly
black and silver makes the bird look as if it is in mourning for itself.
It calls again, a plaintive bruised blues made from just one bent note,
the saddest call of any bird I know. Peter seems not to have noticed
it, getting on with his solitary task of scraping the scales off the fish,
but on the plover's second call he looks up and perfectly imitates it
back to the bird.

As we walk out of the sea with our catch, Peter says he wants his
ashes scattered here on the threshold of the marsh where the sea
runs into it. I think about what he says: a prospect of oblivion beyond oblivion, ashes mixing with mud in the tide, the thoughts of
old age with the memories of childhood, in a wild place empty of
people; but this would also be truly moving, a way of casting off the
separated self and joining loved things, a passage into this world of
constant becoming, breaking waves, rising tides, mullet arriving,
geese commuting overhead, knot coming and going. Gray ash to
brown mud. I turn to look at the Wash behind us. With the tide out,
the great mouth is slack and ribbed, but a few hours from now it will

tighten at the taste of England, as it did yesterday and will again to-morrow.

I want to go out again, too, farther onto the water. But it is not easy to enter this placeless place. A week or so later on another falling sea, I hitch a ride on a fisheries protection boat making one of its regular trips to collect cockles, mussels, and water samples for analysis. We come out of the river Nene and run across the Wash's fifteen miles of open water while the tide lasts and then nose in and out of channels and creeks once it drops.

I feel like a runaway to a sea that isn't a sea. As soon as the water arrives it begins to leave, and by the end of my second tide it is clear, from far out, as it is inshore, that a kind of walking on the water is the condition of human life on the Wash.

To begin with, it is blankly disappointing: empty water and open sky. Then the picture becomes horribly distorted: the screaming yaw of military jets squeeze and split everything at once. The United States Air Force's A10 "tank buster" planes, painted the color of the dry brown mud of the Tigris and the Euphrates, drop their dummies with a *phut*, a flash, and a little puff of smoke into the wet brown mud of the Wash's southern-shore bombing range. They leave behind another of the Wash's mirages, a cloud from nowhere, seen by no one. Wilderness and government guns have a long history: the Wash is like a plain or a moor, officially perceived as an empty and useless nonplace, to be flattened, being blind and deaf to the bombers' hysterics. The planes finish their chores; the mud takes its punishment.

But the Wash runs its own time. The tide makes the day; it is the only clock, and there is nothing else. The brown wave on the brown sea is the mud of the seafloor raised by the tide. The mud of the revealed banks is the settled sea. The same mud that cradles the ordnance and sucks at your legs gives the Wash its teeming life. Within minutes of sailing out onto the wide water, the mud invites us in. We are being taught how the Wash fills as it empties. We seem to be sinking. Peter tells me that fishermen used to say, "We're going down below," meaning they were going out *onto* the Wash. From the boat I watch as land emerges from the sea: mud strips, then bars, then banks appear, like the backs of surfacing animals, an otter, a seal, a whale. Sky and sea both have brown in their blue, as if in readiness to deal with the apparently rising land. A brown streak is visible below the falling water, and then a thin brown seam opens between the sea and the sky, like a desert mirage in reverse, a sliver of land shining in the middle of the wet. A plain becomes a mountain range. Five miles out from the bombed shore the mud begins to rise over us; we sink lower into the world, the horizon looms, the sky shrinks. The Wash comes alive.

First the land and the sea meet in the air. Three gannets, an adult and two immatures, birds of the open sea, and big and bright enough to match it, fly in front of me, drawing their long wings through the wind like turbine blades and moving crazily south toward the green of the Ouse and the great cul-de-sac. Perhaps they have come from the teeming gannetries of fish stew and aggro at Hermaness, where I'd watched them in June. But why are they pulling deeper in toward the land; where are they going? Then three swifts come, like slates skimmed over the falling sea, five miles from the shore, buzzing the gannets, crisscrossing around them, feeding on flying insects in their wake, and as at home out over salt water as they are around the eaves and crannies of the roofs of Boston town. I am thrilled to see them. Would these sea swifts, as brilliantly unplaced and unplaceable as ever, be the last I would see this year?

Swifts are nearly legless and gannets are reluctant walkers; both are ungainly on land and come to it only because they cannot lay their eggs in the air or on the sea. Here over the Wash, with its land that is not land and sea that is not sea, they meet, the brilliant gloss white of the gannets and the black ink of the swifts making broad strokes and fine lines across the gray sky and brown sea, signatures from other worlds.

On the ship's navigation chart in its wheelhouse, the thin edge of true land (only of interest to a sailor in order to be avoided) is marked in yellow; the Wash is colored green for its salt marsh and inner muds, blue for its sand banks, and white for its deepest draughts of water. The chart marks a heart shape of white called the Well; it is the only place that holds open water at low tide. All the mud banks and sandbars have names too, names they carry with them as they drown each tide and shift over time.

While half the crew goes off in a tender to collect samples to be monitored for algal biotoxins known by their acronyms, ASP (amnesic shellfish poison), DSP (diarrheic shellfish poison), and PSP (paralytic shellfish poison), I look out from the deck over the emerging mud-scape to try to identify the banks, bars, and scaups (mussel beds) whose names I had copied from the chart into my notebook. The crew knows the mud by its names, though they don't know who did the naming or when. I think of pirate maps of buried treasure, and of Polonius trying to describe the clouds. The mud names record intimacies and fears, stories of hard-won triumph, and magic words to invoke and appease: Inner Westmark Knock, Gat Sand, Butterwick Low, the Scalp, Styleman's Middle, the Ants, Long Sand, Herring Hill, Black Buoy Sand, Thief Sand, Seal Sand, Old Bell Middle, Stubborn Sand, Breast Sand, Pandora Sand, Bull Dog Sand, Scullridge, Blue Back, Mare Tail, Old South, Inner Sand, the Cots, Sunk Sand.

The tide falls farther and on the next excursion from the boat I join the crew in the tender and we nuzzle into the side of a mud bank. Farther along in the afternoon haze, like a stranding of dying whales, are eight fishing boats strung out on the mud. Settled there by the retreating tide, they are now dozens of yards from the water. For cockle "hand working," the fishermen run their boats up onto a mud bank like this and wait for the tide to drop. Then, in a controlled abandoning of ship, they throw ladders over the side and clamber down onto the mud to rake cockles. Wandering from their boats, their heads lowered, on a bank miles out from solid land, the fishermen look as odd as icebound Arctic explorers, set down on uncertainty. But a good crew on a good site can gather two tons of cockles a day (an eider in the same time eats two kilograms of mussels) and in the salty blur of rising mud and falling water, where there is neither sea nor land, fishing becomes farming.

As I slide into the shallow water at the edge of the mud bank, a tree floats past me, a willow, twelve feet long or so, being carried from the river Welland out to sea. Its thin branches are still green and in leaf; they wave in the wind like a ruined sail, a water-loving tree brokering the meeting of land and sea, like an offering from one to the other, adrift but alive.

As we walk up the mud, Dave explains how you age a cockle by its rings, "just like a tree." He saw a cockle spit muddy water from its buried home in front of us and grubbed it up. It is five years old, with four distinct lines around its clamped white heart and a last dark line, marking this year as its fifth, at its still-growing lips. Dave kneels in the mud, scratching and scrabbling into the brown with his hands, gathering more samples. "This is my office," he says in his softly curved Norfolk accent that seems to rise in the air shaped like the curve of a cockle or the soft bulge of the county's coastline away to the northeast.

Back on the boat we head toward the Norfolk shore, running creeks between the spreading mud. Waders land all about us. Their arrivals always seem like returns, though the water has only just retreated. The world in front looks new made, the sparkling mud with its ferruginous glitter and the dappled water that runs from it, but the sureness of the birds' aerial organization and the accuracy of their touchdowns make you feel they have done it before. There are two hundred bar-tailed godwits and five hundred knot. These birds might have bred and returned already from the Arctic, or like the solitary gray plover that flew over on my previous visit, they could be birds that didn't make it north this year.

In 1995, a twenty-two-year-old knot that had been ringed in its first winter in South Africa was shot on migration in the Bay of Cádiz near Gibraltar. At its death it was returning to Siberia to breed probably for the twenty-first time, and if it had been wintering in the Western Cape all those years it would have flown over 435,000 miles, the equivalent of flying to the moon and back.

Tight flocks of knot thicken above the bank, reflecting down the gray-brown essence of the Wash as they turn. Behind them are leggy gangs of godwits and low, twisting lines of dunlin, like mud splashes tossed into the air. Windsurfing oystercatchers are coming in, too. On the mud the waders mix, but they lift from it and make their flight decisions species by species. Though they are tangled on the shore all the godwits go at once, leaving spaces where they stood among the dunlins. In the sky each species has its own flock shape, but all are part of a flexing cloud, like the silver mullet steering through the water below or the gray-green sueda bushes on the distant marsh edge, touching one another to become a low ridge of wild hedge, or the creamy cockles spitting the slip they have filtered for food, stippling the great whale's tongue of mud.

Into this cloud, people have come, nervous and brazen at once, alone and sticking up into the sky. In 1212 in Cambridgeshire, on the marsh edge of the Wash, King John flew his goshawks at cranes and killed seven. Either the same year or the next, he flew his gyrfalcons in Lincolnshire and took nine more. Then the Wash claimed its bounty. Somewhere in the mud is John's treasure, ditched in October 1216 in flight from the French and now sunk, buried, or washed out to sea.

Other than mud, I cannot see any treasure. More knot wheel and mist over me, and I remember their namesake Cnut, or Canute. A would-be ruler of the waves, he went down to the sea's edge to tame it, but he couldn't even find it and met instead something marvelously unbiddable and variously itself, wet, slippery, awash.

August

In the Hand

> We were all day hunting the wren;
> We were all day hunting the wren;
> The wren so cute and we so cunning.
> He stayed in the bush when we were running.
> WATERFORD WREN SONG

The first living wren I held made me feel I had caught a falling star but, since it was science that had brought us together, I kept quiet and didn't own up to this. We were in a wooden hut at Chew Valley Lake, south of Bristol, on a late August morning and I was banding bags of trapped birds. In any case, it was easy to love the evidence in my hands: the emargination of the sixth primary of chiffchaffs, the blue-gray of the moulted greater coverts of first-year great tits, the reddish iris color of adult dunnocks, the fat keels of sedge warblers. Yet I couldn't get over the sensations I had, with the wren in my hand, of holding first an energetic walnut, as its dark brown dwarf body throbbed with life, and then an earthed meteorite, as its midget roundness drew into itself with a density far beyond its weight.

It is fifteen miles from my bed in Bristol to the hut. We banders have gathered early in the morning and unfurled the eight mist nets down by the lake, which is really a reservoir; my drinking water sits here like oil in a basin. The nets on the reedy shore billow for a moment. When they settle they are almost invisible.

We wait for thirty minutes, then check the nets, four of us walking from the hut across a grass field like a mendicant order setting out in search of offerings. Each of us carries a staff like an odd crozier, a brass hook on the end of the broom handle, which is used to pull the high-slung nets down to our level. In our other hand, for stowing our treasure, we hold a bunch of old faded cloth bags, similar to those used by banks to carry small change but made from scraps of material, tablecloths, thinning curtains—patchworked stuff that makes me think of my grandmothers' aprons.

We reach the nets. Birds hang in them like strange fruit. In front of me are a reed warbler and an English robin. No matter how many times I make that first glance along a mist net and discovered birds hanging there, and no matter how commonplace the birds, their presence remains a magical sight. Trapping birds is like being able to see inside their heads or being able to understand their speech. My fingers prickle. Some old hunter in me is thinking of his supper, then an even older shamanic frisson stirs; we have called down these creatures from their skies and delved into the dark of their migrations. For a moment we will hold them to us, before returning them to their journeys and their secrets.

Skilled ringers are wonderful to watch working a net, untangling the bird silently and fast, like Penelope undoing her weaving day after day, holding back time. The robin and the warbler hang still and quiet upside down, their heads and beaks pointing to the ground, calm but not surrendered, the robin like a chubby thrush, small and ruddy breasted, the warbler like a limp sandy flag. Their flights have been interrupted but not ended.

Like an apprentice surgeon I am keen to approach and touch the bodies in front of me, but I am also frightened of hurting them. I am closer to a reed warbler than I have ever been before. Its eyes pull me in. A reed warbler's eyes are not unusual, a pair of black beads, but being only inches from mine, they seem like the deepest of pools. I wonder what they are seeing. Even this close, birds don't catch our eye, there seems to be no recognition. I feel processed but not known. Their eyes remain quick and keen, unfogged by the experience.

This blank seeing, with its simultaneous inclusion and exclusion, reminds me of the first milky, unfocused looks of my sons more than a decade before and I begin to talk to the reed warbler as I reach into the tangle of the net, in a voice with the same tone and intention I used with my infant boys. The reed warbler squawks and hisses back to me as I shush it and try to hold it steady. Its thin red lips (the inner edge of its beak), orange gape, and tiny arrow-pointed tongue are close and active, jabbing and brave, its dark eye severe. I unpick its pale orange feet first. Its claws have grasped at the mesh and are reluctant to let go of the thing that has trapped them. My shoulders stiffen at the tangle, but the warbler's own elasticity beneath my fingers makes me bolder at brushing the mesh from its feet. I am able to hold both legs and feet at the tips of my fingers of one hand and manage to pull a stitch of the net over the bird's left carpal joint. It wriggles its head back through the square of mesh it had dived into, comes free of the net in my hand, and its mushed feathers, the color of warm dust, return and settle to their reed warbler shape.

I open the first and second fingers of my left hand to a V and slide them around the bird's head, bringing my thumb and other two fingers around its body and over its legs. Its hissing stops and I feel the bird calming just as I become aware of its furious heartbeat, hammering out through its thin chest, terrifyingly fast. In my hand there seems nothing more obvious in the world. I wonder whether it can feel my pulse in return.

I have more birds to extricate. While I have been swooning over the reed warbler, the banders alongside me have already dealt with three or four birds. With my free hand I open the cloth bag and reach into its dark, releasing my fingers and laying the warbler down, quickly withdrawing my hand and sliding it from the bag.

Later the same day I am too slow after finally getting a great tit out of a net with much fussing by me, punctuated by the bird's pointed beak stabs to my hands like a tiny woodpecker. On the way into the bag, it spots its moment, scrabbles up my cuff, and flies bounding away, calling its freedom and annoyance into the trees.

I collect birds in bags from the other ringers and gently sling theirs and mine onto a lanyard around my neck. I look as if I have won twelve cloth medals. I climb the stile to the field, holding the precious swaying cargo to my chest with the same caution with which I took my boys out on our first journeys. Twelve birds hang there, their twelve tiny hurrying pumps answered by my clanging heart. In the hut the bags are gathered on wooden hooks. Sometimes a bird struggles and the bag ripples like a sleepy bat, but mostly they hang still like bunches of old grapes. As the autumn goes on, this effect will be magnified by the wine stains that many of the bags acquire from the blackberry juice that rubs off onto them from the birds' beaks or smears onto them from their droppings. For days after a ringing session, my hands will have blackberry birthmarks, too.

We take down the birds, one bag at a time, and the gentlest of interventions begins. I feel sure I will snap a robin's slight twig of a pale green leg as I tighten the ring around it with a pair of pliers. When all goes well I almost cheer. It is thrilling to have put something of ours onto the bird, some evidence of our embrace that will accompany it back into its world. As I measure and weigh it, passing it from one hand to another, I wonder if anyone will touch this robin ever again. Might someone else from somewhere else take it from a mist net a year from today and rotate the tiny band on its leg to read the number? Would they marvel at the thought, as I would, that another person had already held this bird, which, in the meantime, has been who knows where and has done who knows what?

I am a clumsy beginner and in the hut first a greenfinch escapes from my grip, and then a sedge warbler, both hurtling toward the windows. It's frightening to see how desperate they are to escape.

No amount of crooning will ever make them want to stay. I watch the other banders at work. One is a retired tax inspector and I can't help thinking that his fingers are probing the bird for its secrets; another is a potter, his large plate hands miracles of delicate touch; a third is an electrical engineer, and the bird's functioning technology, how it fits together, is alive to him. My trainer has been banding for nearly fifty years, but the flicker of his first bewitchment still crosses his face. He moves a bird with the fluency of a card dealer or magician, switching it from one hand to the other without it leaving the contact of either.

I, on the other hand, feel like Gulliver in Lilliput. I turn the birds over onto their backs and blow on their bellies to part their feathers along their tiny breastbones, looking first for signs of new feathers, evidence of their molt, and then for their fat and muscle scores. The fat on a sedge warbler makes me think of pâté: beneath the warbler's dark pink skin at the base of its throat, and behind its legs, are tiny discs of creamy tallow that might tempt a gourmand or a starving man. "You could light this one," someone says, looking at another. The fat will allow the sedge warbler to make a rapid long-haul flight from western England across Iberia, North Africa, and the Sahara Desert without further significant feeding.

Almost all of the migrant reed warblers, chiffchaffs, blackcaps, and sedge warblers that we band are young. They hatched this year, just a few weeks ago; they might have forty-seven days where I have years. They may come from the same reed bed or scrub we caught them in, or perhaps from many miles away. Their journey south might already have begun, or they might not previously even have traveled as far as the three hundred yards from the reeds to the ringing hut. All are foundlings who have never been abroad before and have to find their own way to Europe and through Africa. Their parents have already gone, or have abandoned them and are waiting for them to leave before leapfrogging over them, flying farther more quickly.

The banders greet an adult willow warbler taken from a bag.

The old bird is a survivor who has made it away and back again at least once. I scrutinize it for its badges of wisdom and experience. Its plumage differs from that of a first-year bird, and its belly is white, not yellow. It is wonderful to think that the bird in my hand has flown across the Sahara twice at least, part of the loose green cloud of one billion willow warblers that leaves Europe and Asia for Africa every autumn.

I hold the willow warbler, as my colleagues in the hut are holding their birds, in a grip that is a cross between holding a pen and holding someone's hand. Feeling the bird between my fingers and palm makes me happy. I have read it, now I will write with it. While I keep it cradled in my hand, the warbler is like a little grenade, unexploded but with the pin out. When released with the ring, *our* badge, wrapped around its leg, it will go off. This single bird's few grams are a bundled message in the present from the future, unread now, as yet unreadable, but carrying a latent shimmer of meaning that can be deciphered should the shining ring on its right leg be seen again.

In the first decade of the 1800s, around the time of Wordsworth's first version of "The Prelude," John James Audubon tied "light silver threads" around the legs of the pewees (now called eastern phoebes) that bred in his garden in Pennsylvania—the first ringing recorded in America. He wanted to know where they went in the winter and if the same birds returned the next spring. Trapping them and tying them in this way, he made them his, but he also widened his horizon and stretched his mind. The birds carried the past they shared with Audubon back with them. The pewees' air-mail messages went back to Mill Grove, traveling faster and arriving sooner than some of Audubon's letters home to his wife would from his own (later) travels in Europe.

Next to me in the hut, the tax inspector takes down a bag and reaches inside, pulling out a dead chiffchaff, a slight and pale green warbler, its head lolling loosely over his fingers. A cat had killed the bird and the neighboring farmer had brought the tiny corpse to the hut. It was a banded, first-year bird (meaning it had hatched just a few weeks before) and was one of "ours," a bird the station had ringed. No one else would be touching this one. The marriage must be annulled. The ring is taken off its leg; its body, impossibly small in death, is destined for under the hedge.

I am relieved the chiffchaff isn't my bird and didn't die in my hands. A silence hangs for a moment in the room. This is a death intercepted at a time when thousands of young birds are ordinarily dying every night, dropping dead to the ground from a perch or in flight, to be covered in August dew, then eaten or rotting, swallowed one way or another back into the earth. An X is put next to the bird's entry in the banding diary. All the information that can be gathered about it has been; all that remains are rot and statistics.

In the warm August sunshine another chiffchaff is singing from the line of ash trees above the hedge. Chiffchaffs return to song at the end of their summer after their young have fledged, and as I listen to the sweet *chiff-chaff*s, it is obvious that there is no mourning in nature. In a passage in Jane Austen's *Persuasion*, Anne Elliot, feeling cast down and autumnal, comes across a farmer plowing, "counteracting the sweet fruits of poetical despondence, and meaning to have spring again." By their autumn flights south, leaving their dead under hedges, birds could be saying the same thing.

The nets have to be furled. I walk out to them on a final collection round, and as before, birds are hanging in them. Seeing them there, I unroll the last half hour backward, knowing what these birds were doing when I was attending to others elsewhere, seeing into these birds' unseen lives: a song thrush, with its speckled breast in the bottom pocket, its low foraging hops and short flights under bushes stopped; a greenfinch, beady eyed and with blackberry juice dribbling from its thick beak, waylaid on its journey from one bramble top to another; a reed warbler, all monochrome sand, frozen in its flight between reed heads; a gang of blue tits, like thrown splashes of yellow, blue, and green paint held along a net on their joint descent to a seed basket.

We extricate them, take them to our hut, and band them. Then we let them go. Releasing the birds is as moving as taking them from the nets. Our time together ticks down, the business is over, the measurements noted, and the bird reassembles its quintessence in my grip as I reach toward the little wooden tunnel built through the wall of the hut. The way back to their world is down this dark passage. The tunnel ends a few feet from the hawthorn hedge, the chiffchaff's grave. I open the wooden flap, reach in, and unwrap my fingers from around the reed bunting I have just banded, its black-and-white-streaked head, the dark brown body feathers, its white outer tail. I feel it instantly gathering itself in an ecstasy of repulsion to move away from my hand and me and toward the light. That is it. It bursts out of the other end of the tunnel at speed, already flying, calls once as it gains height to clear the hedge, and is gone.

I was nothing to it. The sharpness of the cut, the bird's unminded amnesia, is devastating and welcome at once. I think of the wren landing on my shoulder by the sea in Scotland. How ghastly if news of where the bunting had been was sent around its reed bed, how lucky for both of our species that it isn't. Yet I love the sense of having held a bird that is now deep at the base of a stand of reeds, doing what it must do, able to perch above the water on a stem that could never take my weight, in a place that I cannot possibly go, a bird that I have held, now alive in its marsh. A sedge warbler is on an adjacent reed stem, picking trapped flies from an abandoned spider's web for a last meal at Chew, one more dab of fat for its girdle, before it is drawn upwards by dusk and the swing of stars overhead. The bird lifts into the air above the lake, my ring on its leg catching the tiniest of reflections of moon silver, as the great glitter of the night sky steers the bird south to Africa.

September

Leaving Home

> . . . Peggy dear, the ev'ning's clear
> Thick flies the skimming swallow;
> The sky is blue, the fields in view,
> All fading-green and yellow:
> Come let us stray our gladsome way,
> And view the charms of nature;
> The rustling corn, the fruited thorn,
> And ev'ry happy creature.
>
> ROBERT BURNS

For as long as I have known what migration means I have wanted to be alongside moving birds. Their autumn departures and their spring arrivals have been the timetable in my life. To be deprived of an autumn, its lift, its blow, and all its atmospheres between, held indoors shuttered from the wind and the light or exiled to the seasonless tropics, would be a kind of death for me. My children have been born, my parents have grown old, relationships have been made and foundered and made again, my work has flowered and soured and rallied—all these human adventures are what my life has been built from, yet my years throughout have been rhythmically driven by the step up into spring and the swing away into autumn and the movements of birds through them. Comings and goings. Windfalls.

Fair Isle is halfway between Shetland and Orkney, isolated in the middle of violent seas. A port in the storm of autumn for migrating birds, it is small, two and a half miles long and a mile wide. Most of its shoreline is sea cliffs; inland, its northern half is rough heather moor, its south a patchwork of grass fields and croplands. Seeking migrant birds, seeking migration, I spent two Septembers there, one at the threshold of adulthood, one in my forties.

The onset of autumn lies somewhere between the parties of swallows gathering themselves with conversational twitter on telephone wires in late August and the back-to-school tang of the air in early September. But to witness migration as you do on Fair Isle is something else.

Through yellowing crops and fading grasses, I walk south toward the lighthouse on its southern shore and the heaving salt water beyond. Standing at the schoolhouse, I catch a movement from the corner of my eye that becomes a single swallow flying level with my shoulder, dipping and rising over the fences and stone dikes, pulling its blue-black wings repeatedly through the cooling air. It sweeps up for a moment as it reaches the south harbor and momentarily slows as if punctuating its flight with a comma to mark land's end, before casting off alone into the oceanic blow.

One of the first places in the world where passage birds were logged and studied and where the secrets of migration began to be investigated and unlocked, Fair Isle has been known for its migrants as long as anywhere. As a boy I obsessively studied old reports produced by the island's bird observatory and memorized red-letter days when tiny bundles of feathers took shape as the first recorded example in Western Europe of a warbler from North America or a pipit from eastern Siberia. I dreamed of walking its crofts and cliff edges, its dikes and geos, the sheltered green clefts along the island's coastline. I saved money, and for a month in 1980 before I started university, I went, determined to wipe clear my head of home and school and fill it with birds and movement, before I sat still and read books for three years.

In 2007 I went again to Fair Isle, curious about my youthful self and the sights it had stowed back then. After that first month on the island, I pretty much gave up bird-watching for ten years or so. But the birds never really left me; there is no unknowing the kind of knowledge that birds bring. In my forties, my own autumn coming on, I was keen again. I wondered how the delicious paradox of rejuvenation through departure, the sight of birds being chased by death out of the north, drifts of lost young birds and family parties of others moving south together, would seem to me now that I was someone with children, a mess of relationships, wounds, and fresh starts.

Nowadays many bird-watchers fly to the island, inhale its rarities, and dash away. But there is still a ferry, a boat always called the *Good Shepherd,* that runs south from the Shetland mainland across twenty-five miles of sea to North Haven, the natural harbor near the observatory. I took the three-hour ferry on both trips, eking my savings, but wanting also some sense of distance and journey, some equivalence to put alongside the flights of the birds I hoped to see. On both crossings I was horribly sick and so, twice over, began my education in the awfulness of passage.

On my first visit, some other bird-watchers shipped with me. As the old *Good Shepherd* turned a corner in the sea, from the calm of the Shetland harbor south toward Fair Isle, an invisible hand from above slid the hatch and from then on we were locked in the cabin. All young men, we'd nodded to each other as we had assembled on the quay but had barely spoken. Standing there, hiding behind our binoculars, staring at the sea, each of us had in turn undergone the male trial of cautious performance ("Storm petrel at two o'clock"), asserting our place as bird-watchers without addressing anyone directly. Within minutes of trying to float on the furious sea we were filling sick bags in front of one another. The hand, like a god through cloud (where was *that* good shepherd?), reappeared in the hatch space and we understood, broken and gut-wrenched as we were, that we should pass up our wretched offerings to be tossed overboard, as a defeated gull would disgorge its meal for a skua.

I tried to peer through one of the tiny portholes, clutching at its brass fitting, remembering some advice about keeping the horizon in view to allow it to work as a spirit-level for the stomach. But the sea was everywhere. The boat seemed to be in a permanent trough, with no horizon and certainly no longed-for land anywhere to be seen. The windows were small, I realized, because the sea wanted to come into the cabin and kill us. Nothing other than a floating coffin, sealed and dark, had a chance of getting to the other side.

Twenty-seven years on, I was again scourged by the crossing on a new and bigger *Good Shepherd*. In the passenger cabin the portholes were slightly larger than previously. They made me feel optimistic about the voyage. Then I noticed the seat belts.

As I had waited for the boat I had spotted a huge rusting iron urn, three feet across, held in a stone harness over the remains of a brazier at the edge of the sea. It looked like an inverted giant's helmet; Achilles might have worn it. Neil Thompson—the skipper of the *Good Shepherd* today—told me that his great-uncle, fishing after herring, had sailed open boats hundreds of miles from Shetland to within sight of the Faeroes. Ordinary life back then asked for extraordinary efforts. The urn was used on their return to render down herring livers. "They saw a lot of waves," Neil said. I told him about my first trip south to Fair Isle. "It's a bigger boat but it's the same sea," he replied. My stomach knocked at the base of my throat.

We started out and almost immediately the sea came to the boil. Cold but molten lead was being cooked, the color of the pot back on shore. I was to be rendered in it. I had shunned the seat belts, choosing to stand again at the portholes, my watering eyes pleading with the gods and my innards. It was ghastly. Forewarned those years previously, I hadn't eaten breakfast, but still the skua-sea found some vinegary bile to take from my stomach. The boat could not keep sea time. Nor could I. Thousands of years of ships and sailing, human endurance and ingenuity, had nothing to answer the continuous rolling of the sea except repeatedly to smack into it. Hence the seat belts. We saw a lot of waves.

Back on the shore, I had seen a migrant snow bunting huddled near the pier. It prompted a memory. Four years before, on the lava beaches of Surtsey off the coast of Iceland, I had seen breeding males dropping themselves in charming and delicate song flights, like tumbling black-and-white lacy handkerchiefs, onto sea-smoothed boulders of gray-black pumice. I had allowed myself to imagine that the bird in front of me was a grandchild of those Icelandic birds. Also, before boarding the ferry, I had seen a rarity: my first killdeer in Europe, a striking North American grassland wader like an oversized ringed plover, with triple black bands on its neck, throat, and upper breast; a deep orange iris; and a diamond-shaped tail with feathers at its center colored in exquisite warm sand tones. These are the same colors as the dry prairie earth and short late summer grasses of central Ontario in Canada, where I'd seen a killdeer just a few weeks before. This bird—perhaps a relative?—had been blown from there over the whole of the north Atlantic to find the sand of a Shetland sea loch.

Now I was on the same Atlantic that the snow bunting and the killdeer had crossed. Their miraculous journeys had moved me when I thought about them just an hour before. But the knowledge that I was sharing a sea with the birds was useless to me, because I couldn't see any birds or imagine any bird like a snow bunting there. Sea-watching from the sea is very different from standing on a cliff top letting the sea fill your open mind. On the sea, holding on to whatever raft you have—your wings or your good shepherd—is the only thing that matters.

The churn and retch of the engines brought us to the middle of the crossing, to a stretch of water called the Hole, where, free of the currents around the islands north and south, the Atlantic Ocean meets the North Sea full-on. I tried to recall the calmness of the Well in the Wash. I tried to picture the snow bunting's migration, imagining one outside the porthole, but I could only think of the last thing in the world I wanted to think about: that my stomach was a hole, emptied of its ballast and floating horribly close to my mouth.

We arrived after another hour of mind and body warp. Birders who go out to sea looking for pelagic petrels and other seabirds chum for them, dribbling buckets of rotten fish from the back of their boats to attract the birds. I was chumming too but nothing came. At last a dark slice of solid green rose up out of the waves. Fair Isle never looked fairer. I felt I had swum the twenty-five miles and it had taken me twenty-seven years. Staggering from the boat, my queasy lurch giving way to steadying bliss, I understood as never before the meaning of "landfall" and "haven."

I looked up and saw a party of meadow pipits arriving onto the island from the north, from where I came across the sea. They called with what I heard as relief as they wound themselves down to the grass headland at Buness, just above the jetty. Every autumn day the island proves that tiny land birds cross the open sea without a seat belt, a cabin, or a piloting skipper. I wasn't able to see them from the boat but there were birds there, above me and alongside me: migrant thrushes, buntings, warblers, wagtails, part of a vast and scattered flock, heading south, all the way down the entire western seaboard of Europe, every day from August to October.

A few weeks before, many of them had been eggs. Just days ago, in the brief northern summer in Iceland or Scandinavia or farther east, these young birds had left their nests and fledged. Surrounded by the encouraging buzzing of their brothers and sisters and parents, they had learned to fly from that nest to a tree, perhaps a dwarf birch, a willow, or a fir, a few feet away. Their next journey would be on their own and might take them as far as Africa. None of them would have flown over the sea at all prior to launching themselves up into the air, in the dark, one night in early September. They go, and once they go they cannot turn back. "Tracing a memory they did not have / until they set out to remember it" is how W. S. Merwin puts it in his poem "Shore Birds."

What must it be like to be so sensitive to the magnetism of the earth that you are able to taste the iron in the air; to be drawn up into that air as if evaporated; to feel the inching creep of longer nights pushing you away from what you know toward what you don't? What must it be like to hatch from an egg and look up from a nest and know the stars already? As if your paper-thin skull were a planetarium, along with the smooth curve of your late abandoned eggshell and the cup of your nest, too, as if the skies and the stars had pressed their map into everything there is of you.

Migration cannot but seem a mystery. There have been decades of studies from Fair Isle and elsewhere: ringing and radar, tracking and tagging; robins kept in cages flooded with artificial sunshine; pigeons fitted with magnetic skullcaps and opaque contact lenses; Manx shearwaters transported from their nesting burrows in west Wales and tossed back into the air outside St. Mark's in Venice to see if they can make it home. But migration eludes us more than ever. We begin to be able to explain it, but because in our time we feel it less and less, we cannot grasp it. We couldn't do it and we cannot fathom how birds can. Its reality is as hard to hold as any idea of it. Try to comprehend what the earth is busy with in autumn and spring, in its exchange of light and feathers between the north and the south: on this day, say the third of September, there will be 45 million swallows in the air on their way out of Europe. We are in the middle of it, they fly right through us, but we hardly notice.

If we feel migration at all, it flickers in our imaginations like a story from the beginning of the world. If its rhythms and energies still flow, they do so like the sunken memories of folktales and nursery rhymes. Migration's ghost lies deep in our minds and bodies, a shadowy thread running from the earth to the sky and back to the earth. For most of us nowadays, any connection with it is at best muted. On Fair Isle it was brightly shown to me. We are a mobile species, people who have moved over the waters, who have stepped off the land and taken to the sea. Migrations are hard and risky, but migrants are at home in the world and move between homes in ways that no amount of country retreats or air travel can replicate. We once moved with the sun like swallows. And to witness them doing that is to realize what we have lost.

Hollowed by the sea and bleary but safe on land once more, my nausea rolled back into the mucus-colored surf thrashing the jetty at North Haven. My stomach settled with every step, and it seemed abundantly clear, as I carried my bags up to the bird observatory, that walking and not sailing or flying is what we should do. How wonderful our way of movement felt after the crossing, my easy mastery of it and the truth it offered about the world. The horizon that had been either invisible or mobile for the last three hours solidified in front of me: the brown hills of the north and the softer green south. For the next weeks this was to be my place; my migrations were to be local and on foot. Walking is what I did. I wore Wellingtons and trudged the island every day. Once, at the end of my second stay, while hurrying to see a newly found vagrant, I was offered a lift in the bird observatory's car. I clambered in, and as we zoomed south at a terrifying 20 mph, I felt like a Manx shearwater in a wicker basket being flown to Venice.

My life on the island, like that of the birds I had come to watch, was determined by the weather. All arrivals and departures depended upon it. Rock was beneath me and water surrounded me, but the air decided everything. For two Septembers of my life I was a connoisseur of the wind: its swing around the compass watered my eyes and after a single day of walking the island's paths, I learned by charting my tears' coursing progress down my cheeks to plot the wind's direction and strength—its salted northerly gales, the sweeter westerlies, and if I was walking south and my left eye wept harder than my right, I knew the wind came from the east, the best quarter for a fall of birds.

My first trip had been blessed with easterlies. On one day my nineteen-year-old body stepped back toward some ancient knowledge. My cold left cheek made me hopeful and my ears showed off their barometric skills, registering some thickening isobars from a deepening low. I felt the air twist and darken and detected movement above and behind me, and I turned to find myself in the middle of a cloud of birds condensing out of the sky and descending all around toward the rough fields and croft gardens. They were thrushes—redwings, robins, ring ouzels—and all heading south from their northern breeding grounds, the ouzels on their way all the way to a warmer winter around the Mediterranean, the redwings and robins to us, a milder British winter from their northern breeding grounds.

Watching these birds pour earthward out of the sky, breaking from it, was exhilarating: migration was visible, had a shape, and was happening in step with me. I too was on my way, moving from home, and I was in the midst of the passage of these birds that were doing the same thing. I knew I couldn't reach across the gap that separated us, but this didn't diminish the thrilling sense of community at that moment. Flying kept them from me and our separation—so obvious as the flock tumbled down, sown from the sky like rain—was both saddening and enriching; the impossibility of joining the birds made being tangled up with them even more precious.

The loose flock swirled around me: the wall ahead was dotted with birds that had pulled up after hours in the air, like galloping horses coming to a halt. Here was the planet at work. The birds were exhausted, they wanted to rest, preen, and feed, but their eyes were bright and intensely alive; they had brought something from the sky down with them, an imprint, a picture, or a reflection of the whole globe beneath them, and it was hard not to applaud their arrival. The male ring ouzel in front of me was beautiful, like a dusty blackbird with a gorget of white across its breast. Its bright round eye and sharp black beak pointed back up into the air from where it had just dropped and made it look like the figurehead on the prow of an old sailing boat. Movement was its life, and even as it rested on the stone wall, I felt how the earth is turning us all.

I had seen my first ring ouzel ten years previously at the other end of the bird's year, on one of my first proper bird-watching trips. I was eight. My father, who shared my new interest in birds (within months I was in the grip of an obsession that left him miles behind), had arranged a permit to allow us to visit Beddington Lane Sewage Farm a few miles from our home in Surrey. One of our first visits had been on an early April morning. It was cold and the sky was miserable, but the grass on the banks around the sludge pools seemed incredibly green. The spring was coming.

New to all this, we didn't quite know what to do; we walked the banks and peered at the settling ponds of sewage, spotting a few mallard. A kestrel pulled away toward a factory tower. Pied wag-tails, black-and-white feather balls with sticklike tails, bounced along in front of us. We walked on and met another bird-watcher. Not as steeped in spurious protocol as I was to become (witness my caution in 1980 on the *Good Shepherd*), we asked him what he had seen. He said a pair of ring ouzels were on one of the far banks. At this stage of my bird-watching life the only thing I knew about ring ouzels was that I hadn't seen one before. But being beginners we accepted their arrival as we would the appearance of any bird in our path, with grateful bemusement, and didn't hurry to look for them. We carried on scanning the sludge.

Then, jumping through the long green grass on a bank was the male ouzel, with the female not far behind. I peered through my binoculars and the male raised its head, and its bright white breast band simultaneously drew me toward it and pinned me to the spot. It was as if a spell had been cast between us. I identified the ouzel and fell in love at once. It was a rare bird, but it wasn't its rareness that hooked me; it was the sharp tang it gave off of having come from somewhere else. I looked at it and felt I could see its past, its life in the places it had been, so far from the sewage farm and the bank of green grass.

That evening when we got home I read all I could about ring ouzels and felt the thread still taut between us as I discovered how in all likelihood that same night, as I was reading about them, the birds would lift from that grassy bank up into dark air and plow the skies northward toward their summer homes in the uplands of Britain or perhaps to higher latitudes across the North Sea in Scandinavia. They were already on their way to becoming one of my favorite birds; from then on, down to today, I could never see too many ring ouzels.

On Fair Isle ten years later, at the other end of my childhood, there was another ring ouzel in transit from one of its places and going to another, but stopping briefly alongside me, once again casting its scintillating thread. More birds slipped down from the sky ahead of us both, and time turned over again: the wonder of the moment became part of a story of movement; the birds had come from somewhere and they were going on somewhere else. The ring ouzel on the wall turned away from me, flicked its long wings, and flew on.

I walked farther south down the island, trembling with happiness. I saw many other birds in my month on Fair Isle, far rarer ones: I found a woodchat shrike, a streaky youngster with a grown-up hooked bill, that should have been on the other side of Europe; I became adept at listening for the soft calls of common rosefinches, the blandest of birds with the beadiest of eyes; I knew where a buff-breasted sandpiper blown from North America roosted. One night something drew me out of the observatory and I saw for a few moments a green flickering along the edge of the northern sky, the shaking curtain of the aurora borealis: the north electrified and harrowing the birds south. Another day I watched the islanders gather at South Harbor to collect the flotsam of a tide. A ship had lost a container overboard that had split to release its contents. Training shoes and lumber were expected. These were charged days and nights, but after the ring ouzels and their ruddy train of robins and redwings, I didn't really need anything else. I'd had an audience with nature's everyday sublime and was ready to leave home and make my own tracks through the world.

For more than a week in September 2007, the wind stayed in the west, the autumn refused to be autumn, and very few migrants arrived. I continued to walk the island, being tutored in patience, a pilgrim waiting for revelations. My devotions were simple. My feet remembered stiles and paths from all those years before as I moved from the rough heather moor of the north to the crofted south. The cliff edges and offshore stacks have a lovely string of names, the Reevas, the Geos, the Kames. I stopped at every altar and shrine, the tiny stand of pale oats at Shirva, the nettle bed at Pund, the vegetable patch at Quoy, a miniature plantation of stunted sycamores and dissolute pines, the squelching bog beyond the post office. These sacred groves and holy wells are where clouds of migrant birds have previously fed and recuperated, where amazing rarities have taken shelter and been recognized. But this autumn such sancturies fed the resident twites and house sparrows and shone in the warm sun, the grasses and leaves bending lightly beneath the creamy mild air from deep in the Atlantic.

At last the wind changed. First it blew so hard from the southwest that it seemed to exhaust itself. There was a storm. A terrifying sea built up with slabs of muscled water. A huge swath of sky thickened to a metal plate that slid over the island. Meanwhile the sea was running its own foundry and seemed to rear upward. It looked solid and yet able to pour itself somehow into the sky. As the rain broke from the clouds the horizon came toward me with terrific speed, the sea and sky no longer separate. The ocean I had seen from the porthole of the *Good Shepherd* was nothing compared to this—here it was utterly opaque, a thick and malign overcoat made of lead the size of a barn door two inches from my face, dense and deadly and very old. A doomsday sea, too frightening to watch through binoculars.

I ran away and hid in the island's phone box. As I sheltered I thought of what I could least imagine doing at that moment: being launched from a hundred feet up with my arms pinned to my sides to dive headlong into this sea, a gannet's life; or trying to fly through the storm, unable to avoid the lashing wet or the ghastly tar roll of the sea, struggling headlong into a wall of wind, a swallow's.

I walked back toward the observatory. The rain came over me

like a cruel jazz solo played on a hundred drums. It stung. In the fields the sheep all turned northeast in unison and then stood stockstill. I was soaked. On Buness, the wind pushed a waterfall uphill to become a waterspout. Spating streams rushed over the sodden grass to the cliff edge, but the wind was stronger than gravity and forced the water back up into the air to fountain in a great arc, one hundred feet of sparkle through the gray, a blowing whale's back made from a cliff top. It was beautiful but I cursed and dripped through the blasting, hectoring weather. Through the spray came twenty-four black-and-white snow buntings, rising from the sodden grass ahead of me, birds that had made it over the sea but were forced to drop to the ground, to bivouac against tiny rocks like Arctic soldiers in unseasonal weather.

The next day everything had changed. The storm or *gyndagooster* (one of the many Shetland words for the wind and sea I had learned when hiding from the weather in the bird observatory) was spent and birds came in behind it. Often the best birds—if that is what the rarest birds are—come out of weather like this, ruined and wrecked, blown off course, moribund, wrong. And this is what happened. The wind dropped to nothing. A thin high sky, the color and texture of a much-worn, much-washed blue shirt, lifted my head and seemed to pull the whole island south and east. It was patterned with cloud that was being blown from different quarters at different heights. Some upper atmosphere realignment was under way. The autumn, made in the east, came to the western edge of Europe.

It started with geese. The sea had settled, long calming waves (called *da mother* waves in Shetland) rolled in. The day grew sweeter. After a storm that made me believe in hell, heaven was advertised; the sea grew so calm it seemed loving. I watched oceangoing fulmars fall asleep on it, bobbing in its lightest of swells, their smoky-gray heads swiveled behind them and tucked beneath a wing.

Having walked south as usual, in the cemetery by the South Light I looked for warblers and read the gravestones and remembered the storm: James Wilson, "who died from exposure in a boat"

September 3, 1897, aged twenty years; Stewart Stout, "accidentally drowned" February 11, 1946, aged thirty-five; Karl Arne Emanuel Sundblom, "who perished on a raft through the sinking of S. S. Signe of Mariehamn off Fair Isle" on April 2, 1940, aged twenty. Against the back wall of the graveyard is a small metal cross, a memorial to the Spaniards of the armada ship, *El Gran Grifón,* that tried and failed to get back to Spain around the north coast of Scotland and that was wrecked at Stroms Hellier on the east coast of the island on September 28, 1588.

I lay on my back in a grassy *naust* near the cemetery to stare at the beautiful sky. *Naust*s are boat-shaped casts, small dry docks cut from the turf where the islanders used to haul their *yoals,* or fishing boats. Lying in one reminded me of the bumps and dips of the fields on Bardsey off the Welsh coast, where you can lie looking up at the sky for birds and feel yourself cradled by the dead bodies beneath you of the many pilgrims who journeyed to the island and died there. I remembered Coleridge too, walking in the Lake District in June 1801 and finding "a hollow place in the rock like a Coffin— exactly my own Length—there I lay and slept. It was quite soft."

I was nestling in the *naust,* a perfect combination of a nest and a roost, warmed by the sun, out of the breeze, and feeling drowsy, when overhead I heard the beginnings of the distinctive yapping of barnacle geese. Out of the sky they came, a family party of ten birds, talking their soft and conversational dog speech, shepherding themselves south. They dipped toward me but didn't stop; I could see their black necks and white faces leaning down to look at the island, but they could also see Orkney from up there, off to the south, and the air was mild and good ahead of them and on they went.

Wherever they came from, it was as if they had opened the door of the sky. Almost the next bird I saw was a redstart on a stone wall next to the red barn at Kennaby, the same gorgeous rust color pulsing between the bird and the barn, making my day, as redstarts always do, and confirming its place as my favorite of all birds.

The redstart has all I need in a bird. Every one I see has a bloom on it from the times when we are not together: it migrates; it breeds amongst the freshest, greenest leaves in the sweetest oak valleys of western Britain; it winters in Africa, in that ideal landscape of gently undulating grassland with scattered trees; it is shaped like a small, perfect thrush; it perches with an upright alertness; its song is thin but fine and beautifully apt, like the first glances of sun after rain; both male and female are exquisite to look at; males have a slate-gray back and wings, a warm red breast, a deep black throat, and a brilliant narrow white fringe running around their face above their eyes; females are a lovely soft brown; and best of all both have a warm orange-red tail that shimmers continuously. Nobody really knows why. The moving tail and its redness are commonly that which first draw your attention. The blurry red seems to hang in the air in front of you, as if alive independently of the tail and the bird. It's what Gerard Manley Hopkins would call the redstart's instress, or a birder its jizz, and it is beautiful. This redstart was quivering its tail on the worn carpet of lichen that covered the top of the stone wall at Kennaby. A month from then it would do it from an acacia branch in the savannah belt south of the Sahara in western Africa. The next spring it would do it at the edge of its breeding hole in a tree fresh in leaf anywhere from Spain to the far north of Russia.

After that fire-tail, my last two days on Fair Isle were as perfect as they could be. I had loved the resident wrens and the starlings, the seabirds, the gannets and fulmars, but now the island came to life in a different way, as earth, not sea; not sky, but earth, *our* place that land birds need, that we share with them, where they stop and rest and live among us for a while. The observatory emptied. All the bird-watchers combed southward. The *chack* of the first fieldfare of the winter came close behind the last swallow of the summer, and we were in the thick of it.

There was sunshine and a soft easterly wind. It wasn't a Siberian wind icing the west, but it was more than enough to start the birds off. Where had they been waiting, what anteroom in the skies had they been held in, that they might respond to the slightest change of wind direction within hours? Or had they been able to cross the North Sea from the Scandinavian shore in that time? Perhaps, but these arrivals seemed miraculous. I pished, sucking air through my teeth, chacking and tutting, and a bluethroat appeared from behind a pigsty, a beautiful sort-of-redstart for me, almost equally good, its next year's blue intimated in its delicate pinpricked necklace. All around me redwings lowered themselves from the sky ("wind thrush" is an old name for the birds) and I remembered my first visit to the island and the descending thrushes of September 1980.

On that day—September 23—when ring ouzels dropped around me, 120 were recorded; on the same day blackbirds dotted the island's field like rooks and there were 70 redstarts, 50 blackcaps, 300 siskins, 400 redwings, and 550 robins. The numbers were smaller over the two magic days of 2007, when I saw only one ring ouzel and that single redstart, but the island was still strewn with birds: there were thrushes all up the cliffs, robins along the stone walls. A goldcrest like a northern European hummingbird fluttered at a leaf edge, having made it this far on tiny wings and without a woodcock to ride on. Then, typically for Fair Isle, though there were relatively few birds, some of the island's vagrant specialties appeared and complicated the magic.

Along a fence at the cliff edge at Klinger's Geo, dancing before my eyes was a yellow-browed warbler. Its diminutive green and yellow sherbet fizz of energy seemed to have escaped from the lush shelter of the cleft of the *geo* just below. *"Geo"* is the Norn (the old Shetland medieval Norse language) word for a narrow inlet. The cliffs of the island are indented with them. There are no substantial trees on Fair Isle and the *geo*s with their green and moist skirts of moss and lichen and ferns and their insect-rich warmth are the best place for bird repairs. Yellow-browed warblers breed no nearer than the Siberian taiga. The bird in front of me, almost certainly only a few weeks out of its nest, should have flown in the opposite direction to winter in open deciduous forests anywhere from Nepal to the Malay Peninsula.

In the first decade of the twentieth century, the early student of migration William Eagle Clarke was among the first to suspect and then prove that Fair Isle was a magnet for migrants. He published his findings in two volumes called *Studies in Bird Migration* in 1912. He visited the island several times and went searching the cliffs and *geo*s with his shotgun and some crofter assistants from the Stout family. Think of the scene in September 1908, the men perched and clinging to the *geo* just like a creeping yellow-browed warbler, a bird Eagle Clarke described as "a favourite with me." Then a bang that is lost in the wind and sea roar and a tiny body somehow collected or a corpse out of reach falling to the waves below.

The yellow-browed warbler I saw wasn't going to be shot but it had made a mistake, and it is probable that no amount of nurture on Fair Isle could truly rescue it. Vagrancy is a death sentence. Almost all of the rarities that arrive on the island (and almost all vagrants anywhere) will have the same fate. They are wonderful treasures from far away that we cannot keep and cannot save. There is very little evidence that vagrant birds reorient themselves and correct their journeys. It seems most likely that the yellow-browed warbler, having gone southwest where it should have gone southeast, would continue in this aberrant direction and fly on west out over an ocean that has no refuges, no green skirts, for thousands of miles. That would be the end of it. It would soon be homeless. I was watching a lost child at death's door.

The day ended with a lanceolated warbler in an isolated patch of oats right at the sea edge of the southern tip of the island at Skaddan. Close relatives of the grasshopper warbler, lanceolated warblers are just as skulking and hard to see. They are the birds for which Fair Isle remains best known in the birding world. Like the yellow-browed warbler, they are Russian, most breeding east of the Urals across Siberia and wintering to the south. Almost all of the few dozen British records have been from Fair Isle since Eagle Clarke shot the first on the island on September 9, 1908, "as it rose from some rough grass."

Fair Isle in September is still the most reliable place in Europe to see a lanceolated warbler, and a text message bouncing up and down between the sky and the earth told me that one had arrived from equally unfathomable distances and had been found just a mile south of where I was. I knew already that it would be a first-year bird (all those trapped on Fair Isle since 1908 have been), and I knew already that, like all those others, a watery grave was its certain destination. But still—or perhaps *because* I knew all that—I turned and ran toward it in my Wellington boots.

Though it had flown from Siberia, the warbler now preferred to walk, or rather not move at all and walk only if it must. Familiar with this behavior, the warden waited until the whole island of bird-watchers had gathered and then, with a colleague, stepped over the fence at the far end of the oat crop, and the two of them began slowly to drive the bird out. They got closer and closer to us, until a thin strip of oats was all that remained. Surely the bird couldn't be there, but it was. The warden spotted a tiny movement and the warbler crept away from his boot out into the open between two stalks and then back into the yellow-brown tangle. I saw it well for about five seconds. My first and very possibly my last. It moved with the jerky progress of a mouse, a lovely bird full of bright character, like a runt dunnock, severely streaked but with a cocked tail and a beady eye, and just four inches long. I thought of Eagle Clarke's lead shot and the lead sea that waited offshore. I wanted the child warbler to live forever.

I could still see it inching through the oat stems when I heard the whales. Our backs were to the sea but all of us gathered there turned as one. A family of five killer whales, a bull, a cow, and their three calves, was surfacing and blowing only a hundred feet offshore. It is possible these were the same whales that I had seen on Noss. The line of bird-watchers gasped as if we were all breathing in what the whales had breathed out. The bull's huge black dorsal fin seemed to fill the sky like a huge pirate flag, and the brilliant white of the whales' flanks effervesced in the green sea. As one rolled under, another was surfacing, so a lovely serpentine ripple plied the calm, turning the family into a single beast, inseparable then as they would remain for life.

Another year, another day in September, I was crossing the Bay of Biscay on a ferry and had seen storm petrels. We were straining our eyes looking for spouting whales miles out across the silvery water toward the horizon. The wind was chasing us hard south and attending to the sea's surface with diligence. It picked at acres of water, and as it rose in strength it seemed to go back over the sea to work on smaller and smaller patches until every square inch had its drama. I thought of Kafka's vignette on the sea god: "Poseidon sat at his desk, going over the accounts." There was so much sea and each roll or wave was made up of these inches of water, each with its own ridges and furrows, mountains and valleys, wrinkles and frowns. The wind farms the sea in this way and makes a vast extension of the skin of the earth. And through this, barely over this, come flying migrant land birds.

Two miles of water were below me, a distance that is less known and less knowable than its equivalent in any other direction. By comparison, the thin air of birds' lives seems solid and well mapped. Up from the sea's depths, as from a dream, came fin whales; we slid past one another, and their jaws and undersides lit the sea with an otherworldly blue. Though the ferry was bigger, the vastness of the whale was stupendous. Its back broke the surface and it blew, and the sun caught its breath. The whale rolled forward and went under. Far away it surfaced and blew again and then again. Through these fleeting rainbow clouds tiny things appeared in my binoculars and grew to gannets as the ship reached them or they passed it. Then there was a tiny thing that didn't grow and suddenly was upon us, whizzing overhead, and somehow drawn down toward the ferry and to me. I put down my binoculars and turned my head and two inches away from my face, perched on my left shoulder, was a willow warbler.

It seems unlikely that twice in my life while watching whales a tiny bird should have landed on my shoulder. Yet it happened. To see a land bird at sea is one way to feel how the earth is made and how it can fit a whale and a warbler together. The warbler weighs less than an ounce and has come from who-knows-where to the north; perhaps it was the bird that sang along the railway cutting at the back of my house this spring and lifted my journeys to work; perhaps it had passed through our trapping area at Chew Valley Lake, missing the nets and my grip by inches; perhaps it was a Norwegian bird, the descendant of the three or four in a gully that I had helped flush into a funnel trap on Fair Isle nearly thirty years before; perhaps it was just a few weeks old and had never seen the sea nor a man apparently standing on it. Its washed green back was the same color as my coat; for a moment we were a bleached and shrunken version of Long John Silver and his parrot, sheltering in the lee of the ferry's funnel. A second later either the bird realized where it had landed or the wind whipped it from my shoulder, and it let itself be lifted into the air, out to the side of the boat, south over the whales and over the sea.

October

Cage

Hope, Joy, Youth, Peace, Rest, Life, Dust, Ashes, Waste, Want, Ruin,
Despair, Madness, Death, Cunning, Folly, Words, Wigs, Rags, Sheepskin,
Plunder, Precedent, Jargon, Gammon, and Spinach
MISS FLITE'S BIRDS IN CHARLES DICKENS'S *Bleak House*

No, no, no, no! Come, let's away to prison.
We two alone will sing like birds i' th' cage.
SHAKESPEARE, *King Lear*

At Troy in western Turkey, you can see and feel the Asiatic
steppe reaching the sea. The rolling open land ripples one last
time just behind the shore where the Greeks landed their fleet and
made their camp. On this ridge behind the beach is a huge mound;
Achilles might be buried there. From the top you can look far out
into the silvered Aegean, but the view inland is far more arresting.
This is the view that the Greeks would have had looking toward
Troy. In the dusty lemon light, the ruins of the city lie far off to the
northeast, just visible to the naked eye, raised on a promontory on
the other side of the plain. Between Achilles' tomb and the city he
brought down the country undulates into some of the most beautiful, benign miles—with or without the *Iliad*—I have ever seen. The
Trojan plain has wide arable fields, rough grasslands, smaller mead-
ows, and orchards with scattered oaks sown naturally between them

and, at the Scamander, the flash of river silver. Shade and light and shelter and sustenance are woven into the plain.

Trucks on dirt roads throw up dust, and these moving clouds catch the sun and flicker with specks of mica. Homer describes the advancing Greek army becoming visible to the Trojans in the same way. In October the fields are earth brown, scratched with stubble, and the grasslands are pale citrine yellow. Everything has dried. Most things have died. Tall silver-gray thistles are snapped and broken across the plain in sharp-angled agonies like a dead army on barbed wire. The skin of the earth has been stretched and pulled tight like cracked lips in the sun and the wind. Olive trees shimmer and the oaks—the local Trojan oaks—are a dusty hard green and rattle with their bearded acorns in the endless wind that blows from the land, as if from the past, a wind that has crossed grass all its life and that has never tasted the sea.

The miles of open country below Achilles' tomb seemed made for the *Iliad* and the poem could be read in the plain. I sit for a while with my microphone switched on, sucking the dry wind into my tape recorder. As I listen to it coming and coming and coming, I watch Troy through my binoculars and pick up a swallow flying out of the wind through the nearest oaks, a single bird on the point of leaving the land, like the swallow on Fair Isle. As if all the air of Turkey and of central Asia were at its tail, it flies fast and committed, knowing where it needs to go. The tomb seems a way marker. It flies from the trees over the dead grasses of the mound and right over me, two feet above my head, forcing me to swivel sharply to watch it streaming on over the beach and then out over the sea, flying at the height of Achilles' tomb south toward Africa. This swallow then, not the one on Fair Isle, will surely be my last of the year.

I walk off the mound and the next bird I see is a redstart. A drab first-winter bird, it calls once from a fig tree—which is still fresh green and humming with fruit-gorged insects—and flies out ahead of me flashing, as ever, its irresistible trembling tail. Its red is as warm as the sherds of terra-cotta pots that were crushed into the track beneath my feet. It dives into the crown of an oak, and that's the last I see of it. Goats have come around me and are nibbling at my microphone cables and I have to move on. The next bird I see I think at first is the same redstart, but it is a robin. Hidden in some tamarisks on the beach it *ticks*. Like the swallow, the redstart is on passage south to Africa. The ticking robin has come from the north, perhaps from northern Turkey, or from Russia or Ukraine across the Black Sea. It will probably stay on the beach in these bushes for the winter.

The rusty red of the robin's bib and the rusty red of the redstart's tail remind me that Aristotle thought that the summer redstarts of Greece turned into its winter robins. He called restarts "red tails" and the bird's scientific name, *Phoenicurus phoenicurus,* means "red tail red tail," as if the name must itself quiver. Aristotle identified the similarities of the two species and noted that redstarts departed at about the time that robins arrived. The robin ticks on, bringing the winter around, sending the redstart off, proving Aristotle wrong, proving him right.

On May 10, 1940, John Buxton, a soldier, was captured by the Germans at Hemnes in Norway. He was an English teacher at Oxford and a poet, but he was also a serious bird man. He spent five years, the rest of the war, in south Germany in prisoner-of-war camps. At Laufen (Oflag VII-C) and later at Eichstätt (Oflag VII-B—close to the limestone quarry where the first archaeopteryx remains had been found), he watched birds that came through the camp wire to breed.

> In the summer of 1940, lying in the sun near a Bavarian river, I saw a family of redstarts, unconcerned in the affairs of our skeletal multitude, going about their ways in cherry and chestnut trees. I made no notes then (for I had no paper), but when the next spring came, and with it, on a day of snow, the first returning redstarts, I determined that these birds should be my study for most of the hours I might spend out of doors.

In 1950 Buxton published (as part of the celebrated New Naturalist series of books) a monograph based on his wartime studies of the bird. It is called, simply, *The Redstart*. I think it is one of the finest bird books. I also think it is no accident that what is one of the most poetic works of ornithology should be about the redstart. After the war Buxton returned to Oxford. It is reported that he "owned the best handwriting of the New Naturalist authors." He died in 1989.

The redstarts in Bavaria shared their summer home with two thousand prisoners of war. Buxton recruited other prisoners to help with his observations. Remarkably there were several bird-watchers incarcerated with him: one, George Waterston, founded the bird observatory at Fair Isle after he was released, and another, Peter Conder, later became the director of the Royal Society for the Protection of Birds.

The prisoners weren't allowed outside before six thirty in the morning, so Buxton missed the two hours of spring dawn bird activity. Yet the redstarts he studied must be candidates for the most-watched individual birds ever. In 1943 one pair was observed for 850 hours over three months. For days at a time the birds were in view and the men watched them as continuously as their German guards allowed. Buxton's book contains remarkable detail and his account remains the key reference on the bird to this day.

His ornithology was good, but what makes Buxton's book so unusual and distinguished is his beautifully expressed humility in the face of what he sees. Before the book has gotten under way he is saying he hasn't really written a work of natural history at all: "These redstarts . . . I loved for their own sake and not for the sake of adding to men's knowledge." His modesty, his gently expressed jealousy of the redstarts' freedom, his assertion again and again that the birds might not be doing what he thinks they are—all make for remarkably tender science.

I must be understood to refer only to my redstarts . . . My redstarts? But one of the chief joys of watching them in prison was that they inhabited another world than I; and why should I call them mine? They lived wholly and enviably to themselves, unconcerned in our fatuous politics, without the limitations imposed all about us by our knowledge. They lived only in the moment, without foresight and with memory only of things of immediate practical concern to them—which was their nesting hole, and which their path to it, where lay the boundaries beyond which they would not go; memory also, perhaps, of the way back when their one necessary urgent purpose was done, to the hot sun of Africa.

Buxton's quiet confidence allows him to be serious and playful at the same time. The prison fence and the redstarts coming and going through it proscribed the parameters of his research. Their much longer journeys beyond Buxton's view, on their migrations to and from Germany, become a kind of shadow book to the actual study. He literally cannot go with them, but nor can he scientifically, and this opens his mind. He doesn't stop being a scientist, but as with his studies on the breeding birds that are hedged with doubts, he writes about redstarts in Africa with a curiosity accompanied by a generous shrug. It is as if he is smiling at the unknown: "I wish that some naturalist with a pair of eyes might visit some of these places, or that I had done so myself; but since that has not happened, there seems little point in trying to guess what the redstarts do during these three winter months when they vanish."

The book is full of moments like this of approach without capture. Its overriding image is the freedom of the redstarts. No amount of scientific scrutiny or torn-down prison wire can bring us truly closer to the birds.

THERE IS SCARCELY ANY POINT AT WHICH HIS MIND AND MINE CAN touch: how, then, could I write this book? . . . This book is therefore not objectively about the redstart . . . the bird remains ultimately as unknown and unknowable as ever, and the most that I could hope to achieve was to show to a few of my fellows how much delight is to be had from watching a little bird with its tail constantly aquiver. For

> *He who binds to himself a joy*
> *Does the winged life destroy;*
> *But he who kisses the joy as it flies*
> *Lives in eternity's sun rise.*

Buxton quotes Blake here and elsewhere cites Wordswoth and Racine and Lucretius, Aristophanes and Marvell. Helen Macdonald has studied Buxton's notebooks and papers that are now held in Oxford at the Edward Grey Institute and kindly shared her transcriptions and notes with me. The papers include a letter from Julian Huxley, one of the four editors of the New Naturalist library, who wrote to Buxton in 1948 as he was preparing his book: "I don't quite follow your point about not wishing to pretend to knowledge you haven't got and not wanting to acquire the scientific knowledge to write a scientific treatise on redstarts." A similar query was raised more strongly in a letter to Buxton by James Fisher, another of the series editors: "You have definitely overdone the 'mere naturalist and not a scientist' attitude, almost to the extent of inverted whatnot. Can you please do something mildly about it?"

Huxley and Fisher were great naturalists and understandably allergic to preciosity. And Buxton's book now seems both ahead of and behind the idea of the series he was writing for—ahead because the book makes an account of our separation from nature (metaphorically dramatizing it even, since the watchers were prisoners), and it records a yearning to close the gap between us while knowing that no such closure is possible; behind because it appears not to want to know everything, wants indeed the unknowableness of the redstarts to be part of what a redstart is. This is Buxton as romantic poet. He wanted to be an amateur because he knew the etymology of the word, that an amateur is a lover, and that his love for his redstarts lay at the heart of his experience of them. He quietly insists that a telling of the particularities of his relationships with the birds has to be central to his book. He couldn't write in any other way. This is like Gilbert White or Coleridge, but it is also like Ted Hughes or J. A. Baker, author of *The Peregrine*, the hallucinatory account of a winter watching or imagining—no one could or can tell—hunting peregrines along the coast of East Anglia.

Buxton writing about the redstarts' tail flirting captures the richly contrary drives of his watching and thinking. Here is romantic science, poetry-saturated observation, and a sophisticated self-reflexive anthropology that knows that there is no such thing as objective looking. Yet it simultaneously conceals these ground-shaking truths with a modesty and reserve that could help you survive prison and get the permission of your captors to build the best getaway machine you possibly could: a bird to watch and love that came through the wire every day.

Young redstarts, Buxton discovered, start quivering their tails on their tenth or thirteenth day. The only birds that do it are the redstart species and the closely related rock thrushes. Buxton says they would call themselves redstarts if they needed a name, "since it is by this that they recognize one another."

> The action is performed so often that I do not think it can be said to occur in one particular set of circumstances and not in others, and this fact suggests that it is not a form of display, in that it does not represent any emotion or excitement in the bird itself. Rather it is a means of exhibiting the characteristic red tail to other members of the species . . . the movement brings the tail down and then quickly back to the line of the main axis of the body: it may be imitated by holding the tip of a knife-blade on the edge of a table, depressing and then suddenly releasing the handle.

Buxton asks and looks hard at the birds in front of him, but it seems the red tail and its quiver will not give up a meaning beyond what we can see and guess at. Our looking and noting cannot contain it. It is beautiful, it is vivid, it moves, it says "I am here."

> The displaying male, with his bright tail fanned and pressed down on the branch, his rosy body flattened, his black face and white cap thrust towards the hen; with his wings held straight up to show their undersides as he "waves in his plumes the various light"; his wild, darting flight after the act, accompanied by a sweet warbling song as he flies—all combine to make one of the most strikingly lovely scenes I have ever watched in the lives of birds.

By the end of the book Buxton comes close to apologizing for what he has done:

> Perhaps it would have been better to have remained content with what I saw, to have rejoiced alone in this Ariel delicacy and grace which my book, paying too ponderous attention to such human trivialities as statistics and maps and dates and comparisons, has inevitably missed. For what mattered to me is that the redstarts lived, and lived untroubled except by their own necessities, where I could see them.

This is very different bird writing from what passes these days, but it was provocative in those days, too. James Fisher, Buxton's editor, was a great lighthouse and foghorn in postwar bird life. He wrote a monograph on the fulmar and the still unsurpassed *Shell Bird Book*, a vade mecum of the cultural and natural history of British birds. When I got my copy around 1972 it scared me. I thought that there was just too much in the book for me and that I would never be able to know birds as it did. Geoffrey Grigson's extraordinary countryside encyclopedias prompted a similar reaction. Now I think of Grigson and Fisher as the two last men who knew everything. The trouble is, I think, so did they. There is a loftiness and patrician hauteur in Fisher's prose. He marshals his chapters for our benefit like a headmaster and writes with the slightly bored swagger of imperious certainty.

The Shell Bird Book has many words on music and poetry, but its underlying drive is that birds are over there and we are over here and science has explained it all. Some of the best things in the book are also the most troubling for me. Fisher approves of John Clare— who was much less well known and read then than he is nowadays: "of all our major poets he was by far the finest naturalist . . . He wrote of about 147 wild British birds, 145 from his personal observation, 65 of them first county records . . ."

This is exhilarating stuff if you are a bird lover and keen on poems, too. I had never heard of John Clare before I read about him in *The Shell Bird Book*. The thought that a poet was actually good at birds was exciting. In 1972 I was a tentative newcomer to poetry reading and not that experienced a bird-watcher, either. I was uncertain how the two might meet and knew that both were risky pursuits for a young man to own up to. I mostly kept my mouth shut. I will always remember the horror of starting to read the Molesworth books, a popular 1950s series of cartoon-illustrated books set in a boarding school, and finding that the chief nerd was a bird-watcher and a poetry reader.

James Fisher remarking on Clare's sixty-five first Northamptonshire records now seems interesting for two reasons: to think of a poet adding to science (especially something as exacting as a local avifauna) and to think of someone, a scientist (presumably Fisher), working out what the poet has done. But this is where my worry begins. The birds and the poems are being sieved for their facts. James Fisher, I am sure, would have no qualms about that. But John Buxton, with his redstarts, wasn't ready for this empirical reduction.

Nor would I change this mystery of the redstart's life by trying to make it seem familiar. That may be the method of science: I do not know. But the method of poetry (and that is my concern) is to make things familiar be as if they were not familiar, to make this too much loved world more lovely.

Buxton wanted poetry to be able to touch what we carry away of his birds. He wanted poetic insight to be as meaningful to our total knowledge of the redstart as scientific information. He achieved this, in part, by keeping in front of us the individuality of the birds that were in front of him. His concern with specificity and particularities connect him to Clare, perhaps the greatest poet of the particular in English, and he quotes Clare's poem on a redstart's nest. At the end of his book, Buxton writes,

Yet even now, how little I know of these strange creatures that, merrily busking about the trees which shaded us, or perching on the wire that kept us close, delighted us, as they must have delighted the inmates of Belsen, by the very incomprehensibility of their lives. They lightened (if only for a little while)

the weary weight
Of all this unintelligible world

simply by their unconcern in our affairs, and by the beauty and pathos and vivacity of their lives.

I am deeply drawn to this writing—its precision in retreat and its curtailed ambition; its quiddity, which makes it neither lush nor parched; the way it holds itself back while opening itself to the world. Buxton takes arriving at a border at the edge of things and puts it into the center of his life. It isn't the fence (neither the German wire nor the gap between men and birds) that makes him a good neighbor but the manner in which he looks across it. His book would love to reach and make contact, but it knows that it cannot fully hold on to anything.

I think of Buxton as a man of negative capability in Keats's term, "capable of being in uncertainties, Mysteries, doubts, without any irritable reaching after fact & reason." This needn't imply the hackneyed polarities of science and poetry—it could be, as with Buxton (a scientist and a poet), a way of bringing both together and letting each inform the other. Science makes discoveries when it admits to not knowing; poetry endures if it looks hard at real things. Nature writing, if such a thing exists, lives in this territory where science and poetry might meet. It must be made of both; it needs truth and beauty.

Our habits of seeking truths with birds is deeply human. Collection and classification lie at the origins of all thinking and writing on birds, from cave paintings to the *Handbook of the Birds of the World*. Aristotle was doing it with the redstart and the robin and so were the haruspices and augurs and other readers of nature of the *Iliad*, foretelling what was to come by birds' guts or birds' flight. But much nature writing—especially poetry—has been hostile to the tyranny of the notebook and the ringing pliers and has sought other, less measurable truths. In a letter, the poet Edward Thomas wrote, with unusual vehemence, how "the way in which scientific people and their followers are satisfied with *data* in their appalling English disgusts me, & is moreover wrong." John Clare, wrote of the cuckoo:

Artis has one in his collection of stuffd birds but I have not the sufficient scientific curiosity about me to go and take the exact description of its head rump and wings the length of its tail and the breadth from the tips of the extended wings these old bookish descriptions you may find in any natural history if they are of any gratification for my part I love to look on nature with a poetic feeling which magnifys the pleasure I love to see the nightingale in its hazel retreat and the cuckoo hiding in its solitudes of oaken foliage and not to examine their carcasses in glass cases yet naturalists and botanists seem to have no taste for this practical feeling they merely make collections of dryd specimens classing them after

Linnaeus into tribes and familys and there they delight to show them as a sort of ambitious fame with them "a bird in the hand is worth two in the bush" well everyone to his hobby . . . I have no specimens to send you so be as it may you must be content with my descriptions & observations.

Behind these splenetic outbursts is the idea of man as a fallen creature who has acquired knowledge and lost feeling. This is Adam's curse, and we are all descendants of the first name-giver of the animals who traded proximity and connection for separation and classification. Clare and Thomas are central to my mixed life of birds and words, but both seem wrong to me. Yes, we need to guard against the sterilities of knowing too much or allowing knowing to supplant all else, but we must still know. Not knowing our birds keeps us from them—at Reading railway station, yards from industrial units occupied by companies with names like "Flowtech" and "Phoenix," I watched as three red kites, like flying crucifixions battened to the blue, flew just fifty feet above the platform while dozens of people walked unseeing beneath them to their cars. Removed in this way—not looking up—we can only aestheticize nature like Tennyson's eagle ("He clasps the crag with crooked hands"). Indeed, both Thomas and Clare habitually wrote poems that are far subtler than these prose responses, and their writing is characterized by the continual mingling of *data* with *poetic feeling*.

As a young bird-watcher, more interested in nature's writing than in nature writing, I wanted none of this. Books of words about birds that tried to say more than how to tell them apart seemed secondary, something for after the event. It was the event I wanted, and for much of my first decade of birds I lived with raw facts of hard matter: flight lines, the skywritten scribbles of skeins of geese counted, logs of numbers, the hieroglyphs on eggshells labeled, annotations of feathers, and lists—day lists, my back garden list, my Godstone Pond list, my summer migrants arrivals list, my year lists, my life list, even a list of birds heard and seen on the television; I was simply taking down—one bird after another—the way they write themselves onto the world. Nothing in print could answer or match the four tufted ducks on Godstone Pond or the marsh tits on our back garden bird table that I counted week in and week out. My simple notes on these birds formed accidental phenological accounts—immersions in places that started as a list of birds but became a way of feeling nature and landscape and time, the lived life of the earth.

The only nature writing I admitted to liking at this age was exotic. When I was eight or so, Gerald Durrell and the Hal and Roger series of Willard Price (*Lion Adventure, Safari Adventure, African Adventure*) transported me to the savannah and a dream of a future life. From age twelve on, I began to notice that a few poets took birds into their minds and remade them there without the acuity of observation being lost (when it was, I knew the poem failed or the experience was faked). I allowed Ted Hughes and a handful of other poets (I especially liked the *Voices* anthologies that Penguin produced) to cook my nature for me.

I continued to read, but nature writing—the apparently wholesome picnic of seeing and language—tasted mostly off to me; there were only a few books and a few voices that weren't embarrassing, only a few who saw it as I did or more interestingly than I had and who had managed to get it down. Even these successes had the air of marginal literary activity somewhat guiltily carried out by rather apologetic writers. This quality—it will always be both their strength and their vulnerability—they share with John Buxton. I read Kenneth Allsop, his book of collected newspaper columns *In the Country* and his two novels *Adventure Lit Their Star* and *Rare Bird;* J. A. Baker's *The Peregrine* and *The Hill of Summer;* Desmond Nethersole-Thompson's *Highland Birds;* and Richard Mabey's assorted (and happily continuing) writings. One or more of these books saw me to sleep most nights. I memorized whole swaths of some of them; I knew several pages or so of *The Peregrine* by heart and paragraphs of *Highland Birds*. My stories at school were heavily indebted rip-offs.

Though my principal reading matter was still the annual reports of the Surrey Bird Club or, a little later, the monthly bulletins of the Bristol Ornithological Club, I was aware that I was reading more widely than most of my birding friends. Watchers rather than readers surrounded me; if we talked at all while we were out, it seemed best to stick to the known world of names and numbers. I read my favorite bird poems along with the monthly bulletins, but apart from sensing that these two currents met in me, I could not see a way to bring the streams together. To talk about John Clare's poem on a yellowhammer's nest to my friends in the Field Club at school seemed twee or precious, and, a couple of years later, to talk about the accuracy of Clare's observations in a tutorial at university seemed tweedy or pretentious.

My nickname at school was Tweedledee. At university it was Scared Chicken. My fellow students hardly knew of my bird love. Though I joined the Cambridge Bird Club and couldn't stop noticing birds, I cased my binoculars and kept quiet. My friends watched my close-cropped head, my jerky walk, and my general skulk and saw a defeated bird. When I discovered that they called me I was doubly upset. *Gallus gallus domesticus:* a bird that is not really any longer a bird.

Twenty-five years later, in October at Birdlip in Gloucestershire, I went walking with my sons. I like the name of the place. The wood, which wraps the hillside of the western scarp of the Cotswolds, looks down over the Severn Valley. It had been raining and the trees were dripping. Several had their own scarves of cloud they had made, snagging on their autumn branches and drifting slowly out over the valley. There was not much bird lip: a robin *tick*ed and some redwings flew off in front of us *seep*ing as they went. At the crest of the hill the wood gives way to fields, and a yellowhammer called and flew along the hedge. I remembered John Clare's description of the yellowhammer's nest in a poem ("Dead grass, horse hair and downy-headed bents / Tied to dead thistles") and how excited I had been to find this so close to the description in *The Handbook of British Birds*—"built of stalks, bents, and a little moss, lined horse-hair and fine bents"—and how I also loved Clare's note on the yellowhammer's eggs: "a fleshy ash color streaked all over with black crooked lines as if done with a pen and for this it is often called the 'writing lark.'" I told the boys that a yellowhammer had been the first special bird I had noted on my very first proper (rather than back-garden) bird-watching outing aged seven, and how my dad and I had struggled to clinch the identification. At Birdlip we didn't see much else. I *pish*ed at some long-tailed tits as they came around us, and the boys laughed because I was caught in flagrante, sucking and chacking, as another walker crossed our path. We chatted as we wandered back through the waterlogged dusk light, beneath the dripping trees. I tried to explain the sequence of my thoughts and my enthusiasm for spotting ways in which nature has written itself in the world, but the boys were not that interested. Summer holidays and seabirds on cliffs came up and I sensed that Dominic, aged fourteen, was searching for the word "puffin" He tried "toucan," then "pelican," then "mongoose," and finally had to be helped out.

Nowadays I am most interested in and moved by Dominic's kind of nature talk—writing or speaking of birds that doesn't identify itself as "nature writing." In May 1940, the same month John Buxton was captured, the Germans also caught up with Olivier Messiaen. His "Quartet for the End of Time," featuring his musical accounts of a blackbird (played on a clarinet) and a nightingale (a violin), was premiered in Stalag VIII-A in Görlitz in January 1941. Listening to it is like being able to hear with another set of ears, perhaps even with a blackbird's or a nightingale's. Walter Scott extends what nature writing might be, when "out hunting and with some good lines suddenly in his head, [he] brought down a crow, whittled a pen from a feather, and wrote the poem on his jacket in crow's blood". A single letter from D. H. Lawrence, to Lady Cynthia Asquith in November 1915, gives the finest account of autumn in England that I know. Bruno Schultz's stories, from

under the eaves of Drohobycz in old Galicia, of his relatives transformed into condors and other birds mean a lot to a scared chicken.
Yves Klein's photographed leap ("Le Saut dans le Vide") from a wall
in the Rue Gentil-Bernard, Fontenay-aux-Roses, in October 1960 says
things to me about flying (we can't) and falling (we will). The lysergic
shamanism in Ted Hughes's crow poems turns looking at birds into a
kind of bad trip, or a gruesome hangover, that I hadn't realized I'd had.
Brancusi's "Bird in Space" sculptures look to me as a bird feels in my
hand. Joseph Cornell's box constructions, like Central Park in an aerial
view of Manhatan, with their cutout parrots and collaged landscapes
remind me of nature tables at school and my mantelpiece today, a
world salvaged and in retreat, like a nest but also like a coffin. Peter
Szöke, on his bizarre record *The Unknown Music of Birds,* slows down
woodlark song and makes a soloist from the Budapest opera sing it so
beautifully that I weep to hear it. Ian Dury rhymes and associates
"gannet," "Thanet," "prannet," and "Janet" to great comic effect, and
every gannet since I heard "Billericay Dickie" has been marked for me.
Rimbaud kept a list of pigeon names and terms in English that he assembled when he was living in Reading prior to his great escape out of
the European world. It is hard not to hear him among these carrier pigeons, shuffling his own wings in preparation:

> Pigeons: homing—working—fantails
> pearl-eyed tumbler—
> shortfaced—performing tumblers
> trumpeters—squeakers
> blue, red turbits—Jacobins
> baldpates—pearl eyes—tumbles well
> high flying performing tumblers
> splashed—rough legged
> grouse limbed
> black buglers
> saddle back
> over thirty tail feathers

For a few decades through the twentieth century, we weren't sure how admissible this great flock of birds—noted and then re-imagined—could be to our lives. Nature writing was dismissed as being hopelessly elegiac or sentimental, overdone and out of touch. But, though their binoculars might have been packed away, all sorts of observers went on noticing birds.

Just north of Pisa in a prison camp on the other side of the Alps from John Buxton and Olivier Messiaen, Ezra Pound was held after Italian partisans arrested him in May 1945. He had been making pro-Mussolini radio broadcasts. For a time Pound was kept outdoors in a wire cage with minimal protection from the elements. Afraid of losing his mind, he watched birds. Some of his Pisan cantos, the poems he wrote after his confinement, contain his field notes of birds—swallows mostly, I think—seen on distant wires of the camp. Canto 79 has a running list, one of its punctuations, of birds seen:

> with 8 birds on a wire
> or rather on 3 wires
>
> 4 birds on 3 wires, one bird on one
>
> 5 of 'em now on 2;
> on 3; 7 on 4
>
> 2 on 2
>
> 3 on 3
>
> with 6 on 3, swallow-tails

The birds move freely and Pound in his wire cage notes them down. In Canto 82 he sees the settled birds as musical notes on the stave of wires:

8th day of September
    f f

        d

           g
           write the birds in their treble scale
Terreus!   Terreus!

Old birds and old poems come to Pound in prison. The nightingale cries "Terreus!" —Tereus raped Philomela, who, transformed into a nightingale, called his name—and another canto (75) is almost entirely a transcription of the score of the French composer Clément Janequin's "Le Chant des Oiseaux" from around 1500. The song is an extraordinary, delirious imitative round and chorus that creates the sensation of falling through a spring wood at dawn, with one bird chasing another in song, and has you catching again and again at species (nightingales, starlings, cuckoos, blackbirds). It is cool and green and happy and it comes from far away. *"Détoupez vos oreilles,"* bend your ears, the lyric instructs. In Pound—in the midst of poems begun in captivity—this speaks of the double impossibility of capture. The musical score (my description comes from listening to a recording, not interpreting Pound's scribbles) joins his various languages and alphabets and references as another more or less concrete poem that is difficult fully to grasp. And the human mimicry of nature can only be that—an imitation. We can be captured but the birds cannot; they fly onto the wire but are not caged by it. At Metato—the site of Pound's prison camp—nothing remains of the wire or cages today, but when I visited, there were plenty of swallows flying over the maize fields and singing from the wires along the riverbank.

In the early 1960s on the Neringa, or Curonian Spit, in Lithuania (then part of the Soviet Union) an ornithologist, D. S. Lyuleeva, carried out experiments on swallows. Her findings were published in a scientific paper in 1973. To determine the energy costs of birds' flight, she caught swallows, house martins, and swifts near their nests on the spit and released them 40 or 70 kilometres south. Hirundines and swifts feed as they fly, but to ensure they couldn't, their bills were tied shut with a thread passed through their nostrils. Before they were thrown into the air, they were kept for three hours and deprived of food to empty their guts, and some "measure" was taken to prevent them defecating on the wing (similar experiments with tippler pigeons involved their cloacae being "sealed" somehow). Those birds that made it back to their nests were retrapped and had their mass losses calculated and their bills untied. Not all the birds returned. One house martin was found thirty-three hours after release; it had lost a quarter of its body weight. Lyuleeva described what she did and what she saw: "It reacted to shaking and pushing with a feeble fluttering of its wings. Despite the fact that it did not appear excessively emaciated, it perished after some time. This episode suggested that considerable losses of weight, characteristic of swallows deprived of food for a long period of time, induce torpidity and subsequent death." Perhaps I am missing something, but this doesn't seem a major scientific breakthrough—if a bird cannot eat it dies.

In the 1960s, at the time of these experiments, the Baltic shore of Lithuania, Latvia, and Estonia was a military zone, a heavily policed frontier of the USSR, with patrol planes, watchtowers, border guards, and just a few permitted scientists from Leningrad stitching shut the beaks of swallows. The three countries (then republics) were cut from their shore, their people kept from their sea. The edge of the Soviet Union here was soft: thinning pines and dunes flattening to pale golden beaches, the shallow Baltic coming and going with the slightest of tides. Declared out of bounds to ordinary people, the bow-line of sand became a slowly shifting northern desert.

In a decade of concrete and iron, walls and wires, rust and weaponry, on a cold Soviet shore an ornithologist is sewing a thread through the nostrils of a swallow, a bird that had come freely into that pallid spring from its winter in the skies of South Africa, another country then savage towards its own people. It is hard not to see some human envy, unspoken but deep, at work in these experiments. The birds come and go, we are stuck here, let us catch them and tie them to us.

The wall has fallen now; Lyuleeva's paper is still cited in ornithology books; the swallows are still nesting along the Baltic shore. I saw them there one midsummer night—the sun once more stalled along the thin horizon—hawking for insects between the ruins of a watchtower, picking at the flies a crane disturbed as it walked through a meadow of yellow and purple flowers that held the sun like a million light bulbs. Flying just above the flowers and the crane, the swallow sang as it hunted.

Through the Second World War, German ornithologists were studying redstarts at the same time as John Buxton watched his. Breeding birds were ringed in the Magdeburg and Dresden districts in 1944 and recovered in Spain and Portugal the same autumn, giving information about the birds' routes out of Europe. Earlier research on what triggered migration urges in redstarts had been done at the bird observatory on Helgoland, the curious island (British from 1807 until 1890, when it was exchanged with Germany for Zanzibar) off the German coast in the North Sea.

John Buxton's brother-in-law was Ronald Lockley—another great bird man, an island lover, pioneer student of migration, devotee of seabirds (his film on the gannet made with Julian Huxley won an Oscar in 1938), and champion of bird observatories (he set up the first on Skokholm in west Wales in 1933). It was he who posted Manx shearwaters from Pembrokeshire to Venice. In October 1936 Lockley went to Helgoland to watch some autumn passage. He arrived at the same time as a large fall of southbound birds: "song thrushes and robins predominated. We passed one woman scrubbing the roadway—all citizens must keep the road outside their houses spotless, for dirt is forbidden and litter must be jettisoned through a chute into the sea. There were two robins perching, exhausted, upon the woman's head." That day 752 birds were caught in the observatory's four traps. Wrecked migrant birds falling accidentally down rainwater pipes made the drinking water on the island taste gamey. The next day 350 woodcocks were shot (one man alone got sixty) and sold for two shillings each to be exported to Hamburg.

Lockley reports a trip to the cinema on Helgoland where he watched a Nazi newsreel with "goose stepping, and fierce speeches." He also visited the bird observatory and was shown an experiment where the responses of trapped redstarts and robins to daylight and day length were being studied, and birds were being made to want to migrate by being flooded with artificial sunshine.

> Dr Shildmacher showed me his redstarts and robins living in special cages with delicately poised perches, which by means of electrical contacts register the restlessness of the birds at night when the migration fever causes them to vibrate their wings and jump about the perches. Every quiver breaks the inked record which revolves slowly over a time graph, showing the exact second the fever began and ended each night. By this means it is proved that the direct cause of migration is fever or intense physiological rhythm of energy which seizes the bird. No matter if confined in the darkest cage, when the hour comes—usually beginning towards midnight and ending just before dawn—it must expend that energy in violent wing trembling and beating.

The German word for migratory nocturnal restlessness—wing whirring, perch hopping, body turning—is *Zugunruhe*. The term is still used by ornithologists. Wing whirring is what the redstarts do when they have adapted to their cage. It stands in for the flying they have been prevented from doing and has been described as "migration in sitting position." Another experiment involved would-be migrants themselves writing their intentions—inkpads were placed on the floor of their cages and the marks scratched on the papered sides of the cage by the birds' feet revealed their levels of activity and directional preferences.

John Clare had been incarcerated in madhouses for thirteen years when, in 1850, he wrote a few letters in code from Northampton General Lunatic Asylum. His code is relatively easily cracked, though a few ambiguities remain. Clare wrote to several women. Two letters to Hellen Maria Gardiner have been published. In one he writes of wanting to kiss and cuddle her and to take her flower picking and bird nesting. He cannot, he says, because he is trapped in a prison. The code makes his words look chopped up and boxy. On the page his writing could be mistaken for hasty field notes or the semblance of a song transcribed, the best that can be done in the wrong medium. His letter is an escape telegram scratched from a cage—an indication of Clare's directional preferences. The poet who didn't punctuate his lines, whose childhood and youth had been unenclosed, who made nature's mess his ms—his manuscript—is here hedged, edited, and confined. Torn from his right mind, he has been tipped from his nest.

M Dr Hlln Grdnr
Hp t wll n hppnss nd hlth—nd tht r prtt fc kps the rs blm s t ws whn
mt—b th frdls—xxxxx xxx xxx xxx kss's hw s Crln r Sstr—M drst
Hlln Mr hw lng t pt m rm rnd r btfl nck nd r chk nd lps—"Thn drst
Hlln ll lv n mr"—M dr hlln hw shld lk t wlk wth n th bnks f th rvr
& gthr wld flwrs nd hnt brds Nsts—bt hv bn tn rs n prsn—nd cnnt s n
thng f plsr r pstm—Lv nd nj yr slf m dr Hlln Mr nd b rslf fr hr w e
hmd [? tll] m scrl m wn nm—stll m mslf snsbl s ver ws

m drst Hlln Mr | m rs sncrl
Jhn Clr

This Clare's editor, Mark Storey, transcribes as:

My Dear Hellen Gardiner
I hope you are well in happiness and health—and that your pretty face
keep the rosy bloom as it was when I met you by the firdales—xxxxx

*xxx xxx xxx kisses how is Caroline your Sister My dearest Hellen
Maria how I long to put my arm around your beautiful neck and your
cheek and lips—"Then dearest Hellen I'll love you een more"—My
dear Hellen how I should like to walk with you on the banks of the
river & gather wild flowers and hunt birds Nests—but I have been ten
years in prison—and cannot see anything of pleasure or pastime—
Live and enjoy yourself my dear Hellen Maria and be yourself for
here we are homed [? till] I am scarcely my own name—still I am myself sensible as ever I was*

my dearest Hellen Maria | I am yours sincerely

John Clare

Twenty years earlier, in May 1830, Clare had recorded how his
son found a nightingale's nest. He describes it in detail and with an
accuracy that is as good as anything ever published on the nightingale. The precision of his looking and the quality of his thinking
about what he sees make his imprisoned desire to "hnt brds Nsts"
even more upsetting to contemplate.

May 29 My Frederick found today saturday a Nightingales nest in
the bottom of the orchard hedge with 4 eggs in it and tho there is
but one oak tree as I am told in the Lordship she has got some oak
leaves about her nest—in the woods she generally nay always uses
dead oak leaves very plentifully at the bottom or outside of her
nest and seldom or rarely puts any within side but here she had got
dead grass on the outside and a few old oak leaves eaten bare to the
fibres by insects withinside her nest and I never in my life as yet
saw a nightingales nest without oak leaves and I have found a many
and as many as seven one May in Bushy Close and Royce Wood at
Helpstone

In 1850, when he felt himself to be "scrl m wn nm," Clare was
still certain a nest was the best possible haven in the world but he
also knew he was far out at sea.

November

The Gorge

Look through my eyes up
At blue with not anything
We could have ever arranged
Slowly taking place
W. S. GRAHAM

On a cliff top, six elderly men are sitting in a line of low-slung deck chairs. Two tend a kettle boiling water on a camping cooker; others ply their thermos flasks. Between their attentions to their hot drinks, all of them look ahead, as if in the front row of an outdoor theater. Spread and settled in their seats, they are waiting for something to fall from the sky into the gorge in front of them.

The Avon Gorge cuts at the western edge of Bristol like an injury, a three-mile-long, quarter-mile-wide, two-hundred-foot-deep gash through dirty limestone that bleeds brown mud and tidal water. Its plunge is most startling just where the watchers are sitting at the edge of the Downs, an open city park of mown grass and scrub fringed with stolid Victorian villas. There is a last soccer field and then the sudden calamity of a sheer cliff face. I must have walked to the edge there a thousand times, but every time the falling cliff tells me that my journey, however flat and inconsequential to this point, has nonetheless been made across the surface of the earth. For here the surface has been cut open.

The gorge marks the edge of the city. Where I stand and the watchers sit, feral pigeons behave like rock doves, tumbling over the cliff. On the far side of the gorge along the gentler heights of Leigh Woods, wood pigeons break cautiously from the canopy of trees and hug the brown edge. Neither species, urban or suburban, seems keen to cross the open air between.

There are two reasons for this. The gorge is a wind machine. The bash and thrust of air funneled and collected by its cliffs and slopes makes its own wind, separate from the surrounding city's, and all sorts of local peculiarities blow: gusts, eddies, thermals, vacuums, updrafts, whirlpools of air, and aerial maelstroms. The gorge does what it wants with air, and its birds know this. The pigeons cope as best they can; the gorge's jackdaws and ravens are at home in it; but one bird knows best how the machine works and how to turn it up, and it also wants to eat pigeons: the peregrine falcon.

The watchers are waiting for peregrines. A pair has fed and bred in the gorge since the late 1980s. When I moved to Bristol in the mid-1970s there was none here. I hardly ever saw peregrines then. One, a dark arrowhead, had flown fast along a slate mountain ridge high above me in north Wales, as if it had chiseled itself from the cold rock beneath; another I saw for just a few seconds one winter, a heavy falcon flicking past the stinking commercial henhouses in Bridgwater Bay.

For a time it seemed that the peregrine and the henhouse were not so far apart. Close to extinction in Britain, the most powerful falcon was also the most vulnerable. Mastery of the air hadn't been enough to save it from the insidious creep of pesticides up the food chain, concentrating in the falcons' bodies and poisoning them or weakening them so that the females' eggshells became so thin that they were broken by the adult birds as they incubated. (Books like Rachel Carson's *Silent Spring* and Derek Ratcliffe's work on the peregrine, which identified these problems, made for apocalyptic reading.)

I grew up thinking of peregrines as sickly. They were rare and remote things glimpsed in distant or severe places, but I knew them to have been ruined by the malign human idiocies that were contaminating the whole world—chemical weapons, Agent Orange and napalm, a nuclear missile head and a neutron bomb, death delivered by stealth out of the thin air. The magnificent hunter, the apotheosis of the wild, the falcon on the king's gloved fist, was becoming as helpless as a spastic hen, a bird that broke its own eggs.

This was true, but the peregrine in my young mind was also built by J. A. Baker's *The Peregrine*. I read it when I was eleven, it stole into my mind and stayed there, and I then reread it compulsively. At that time it was mostly ignored by birdwatchers; more recently, it has been noticed by poets. The book has been in and out of print several times since it first appeared in 1967. Next to nothing is known about its author. Though steeped in what seems to be real observation, there are times when you feel you might be reading fiction. Over ten seasons of watching, Baker says, he found 619 peregrine kills. He isn't specific about where and when. Might he have made it up?

Baker's prose seems bruised by his encounters with the hunting falcons (and perhaps also by illness—the book feels written in the knowledge of sickness). It takes on purpled tones and at times is too rich, blood-drenched and strewn with corpses, like Ted Hughes's bird poems. But it is also wonderful: a grounded man's love affair with the airborne, a story told by a downed Icarus long after his fall, earthed, haggard and self-loathing, traipsing through marshes, crouching in ditches and lurking on field edges, and looking up before a greater creature.

> At eleven o'clock the tiercel peregrine flew steeply up above the river, arching and shrugging his wings into the gale, dark on the grey clouds racing over. Wild peregrines love the wind, as otters love water. It is their element. Only within it do they truly live. All wild peregrines I have seen have flown longer and higher and further in a gale than at any other time. They avoid it only when bath-

ing or sleeping. The tiercel glided at two hundred feet. Hundreds of birds rose beneath him. The most exciting thing about a hawk is the way it can create life from the still earth by conjuring flocks of birds into the air . . . In long arcs and tangents the hawk drifted slowly higher. From five hundred feet above the brook, without warning, he suddenly fell. He simply stopped, flung his wings up, dived vertically down. He seemed to split in two, his body shooting off like an arrow from the tight-sprung bow of his wings. There was an unholy impetus in his falling, as though he had been hurled from the sky. It was hard to believe, afterwards, that it had happened at all. The best stoops are always like that.

Baker seems to want to lose himself completely, perhaps even to become an object of prey for the birds. At one point he drops to his knees and creeps through frosty leaf-mold toward a peregrine that has just killed a woodcock. He gets within four yards, "but it is too far, a distance as unspannable as a thousand foot crevasse. I drag like a wounded bird, floundering sprawled. He watches me . . ." An even stranger meeting of bird and man occurs when Baker is sheltering in a barn and a wood pigeon, struck by a stooping peregrine, hits the roof and slides down in front of him almost dead. He kills the pigeon and tosses it to the ground, where the peregrine collects its prey.

You can see why some bird men were skeptical about the book's authenticity. The weirdness of *The Peregrine* cannot be overstated, but Baker writes with an extraordinary authority. There are no notes or references. Along with things he has seen or appears to have seen, arcane lore from books is included (there is an aside on the crow catchers of Königsberg) and facts that he might have invented (surplus meat from abandoned peregrine kills helps to support, he says, "tramps and gypsies"). He says long-tailed tits use the feathers from peregrine kills in the construction of their nests: "I have found an unusual concentration of such nests in areas where many kills have been made." This is a unique observation, as far as I know, not supported by any reference anywhere else in the ornithological literature.

The book's heart is hunting, and specifically the peregrine's sky-diving stoop. Baker works over many kills to find some written equivalence to what he has seen. His obsessive nature as both watcher and writer makes the peregrines' daily hunting habits seem obsessive themselves. Obsession, though, is meaningless to the birds. They hunt all the time but have no conception of hunting as an all-consuming activity. But Baker needs the extremity of his driven life as falcon follower to be matched by the bird. Early on he says that the peregrine is "inhibited by a code of behaviour . . . if the code is persistently broken, the hawk is probably sick or insane." Insane? How would we know? This seems more like a veiled announcement about Baker's own code breaking, his stepping away from the human world. If his peregrines seem mad, it is because he has made them so. At times his writing has an iron-cold sadistic brutality: "The odd are always singled out. The albinos, the sick, the deformed, the solitary, the imbecile, the senile, the very young; these are the most vulnerable." He is nature writing's Goya.

I found myself crouching over the kill, like a mantling hawk. My eyes turned quickly about, alert for the walking heads of men. Unconsciously I was imitating the movements of a hawk, as in some primitive ritual: the hunter becoming the thing he hunts. I looked into the wood. In a lair of shadow the peregrine was crouching, watching me, gripping the neck of a dead branch. We live, in these days in the open, the same ecstatic fearful life. We shun men. We hate their suddenly uplifted arms, the insanity of their flailing gestures, their erratic scissoring gait, their aimless stumbling ways, the tombstone whiteness of their faces.

At other times and for much of the book the imagery is of great precision and it anchors the weirdness: a heron trying to land is like "a man descending through the trap-door of a loft and feeling for a ladder with his feet." Baker's best writing is about the flight and fall

of the birds he sees above him. He is wonderfully good at going up
with them, though he is always at pains to stress his—our—earth-
bound life: "Imprisoned by horizons, I envied the hawk his bound-
less prospect of the sky. Hawks live on the curve of the air. Their
globed eyes have never seen the grey flatness of our human vision."

The Peregrine was written at the time when the catastrophe con-
fronting the bird was taking hold, and it ends in malignant crisis and
a sense of the falcon's pressing extinction. At that time the peregrine
was considered globally threatened, and it was thought that it
couldn't survive in our world. It had nowhere else to go; it could not
retreat to the wild places because the pesticides found it wherever it
went. In the book's grimmest moment, a peregrine falls back to
earth, dashed down from the skies like its own prey, stunned by what
has hit it, poisoned by chemicals, and clutching skyward with its yellow claws.

For me this was the greatest and most needed bird book of my
youth. For years I returned to loved and memorized passages and
breathed its poisoned skies with a shared horror at the coming vac-
uum, the air without the birds that give it shape.

In 1973 I moved with my parents and sister from Surrey to Bristol. I was twelve. Things weren't good at home. My father had tried
to relocate his business from Croydon to Bristol, but it went wrong
and he ended up losing his job. He had had an affair, it turned out.
He was sacked. He was also drinking too much. One evening back
in Croydon, not long before, my sister and I had met him after
school and the three of us had gone for a meal near his office. By the
time we left the restaurant he was too boozed to drive, but we had to
get home, fifteen miles away out into the southern edge of suburban
Surrey. He drove. My sister climbed into the back seat of the car and
fell asleep. I sat next to him in the front, wide awake and terrified.

The roads were dark along the wooded hills of the North Downs.
On a previous night ride home my sister had called these "hairy
woods" because of the way in the car lights the tree branches rather

scarily seemed to be plaited together above the road. This night a tawny owl suddenly appeared, dusting the edges of the beams, flying down the meshed tunnel in front of us in its brown field of gathered quiet, showing us the way for a few yards before disappearing into the black curtain of the woods at the side of the road.

The owl knew where to go and my father sobered for a moment. But within a mile he had arrived at the fond rambling stage of his drunkenness, interspersed with gulfs of silence. I was trying to keep him talking. On a back road, out of the woods but still a few miles north of our house, he nodded forward, his head dipping to his chest. There was a parked car in front of us at the side of the road. I watched it float toward us for a second, like the owl in the lights, then reached over and pushed the wheel to the right with a twelve-year-old effort at steering. The jerk of the car woke my father enough for him to finish the maneuver.

We made it back. My mother was waiting for us. I started crying when I saw her sitting on the foot of the stairs in the hallway, watching our bleary and hooded return into the lit world of home. My sister and I went to bed; I fell asleep to the already familiar night music of my parents shouting.

In this state we arrived in Bristol and were immediately stranded. My parents were in crisis. My father couldn't find a new job and had to go on welfare. We had to move to a cheaper home. I watched my family, horrified by what my father had done, terrified that my parents would separate, desperate to escape the corrosive air around them both.

I watched birds. Where people clashed and collided, birds always seemed to know where to go and never bumped into one another. At the time I didn't—couldn't have—put it like this. Then, on my paper route one afternoon, I saw a man jump into the Avon Gorge and I learned as deeply as seemed possible, then and now, that we cannot fly and we must make our escapes in other ways.

I had had a morning paper route but I liked the afternoon one more. It helped fill a drifting hour between the emptiness of school and my cautious return home to our unhappy house. It also took me toward some birds. I cycled with a bag of evening papers over the gorge across the Clifton Suspension Bridge toward the houses of Leigh Woods, where I had to make my deliveries. Jackdaws were almost always out in the middle air of the gorge as I crossed the bridge. It was hard not to think that they loved the testing winds. The birds, grouped in a conversational cluster with their power-gray heads and black bodies like drifting ink and soot, would drop hundreds of feet simply by folding themselves up. Then, flicking open their wings in unison, as if powered from beneath, they would rise up to the handrail of the bridge from below, coming to my eye level and *chack*-ing. Their garrulousness always made me feel lonely: how wonderful to live like a jackdaw, always among your own kind and surrounded by reflections of yourself; to be a single organism and a community at the same time—wasn't that a definition of a family?

I looked beyond the jackdaws at the cliffs like marbled meat on one side of the gorge, the bosomy trees bucking in the wind on the other, and far down to the hemmed-in river below, toiling muddily through the slash it has cut, wanting to be an estuary and to escape to the sea, and meeting a daily silt-thick tide of salt water, wanting to be a river, pushing deep into the heart of the city. Some days a briny tang came up the gorge and I felt I could taste the Atlantic; on others it was a stink of all the rank socks and rotten vegetation of the West Country. I saw oystercatchers on the mud, black, white, and orange like marine buoys, beneath the green oak and ash trees of the riverbank. They piped their seaside yells into the landlocked woodland chorus of songbirds.

On the far side of the bridge in Leigh Woods, the middle of the afternoon was quiet; the place seemed buried into itself. As I cycled up the switchback of the wooded road, I felt I was opening the place up to its evening, bringing something of the jackdaws with me, pushing air into a somnolent world.

The birds helped. There were always birds. On summer afternoons thrush song and blackbird song was just starting up again at four o'clock, spilling beautiful woodwind descants from the trees. On winter dusks, robin ticks and wren flourishes made a percussive crepuscule for the settling wood. The autumn was full of feeding movements of tits and finches; the spring, of unfurling leaves and warblers' sung arrivals.

The woods in the 1970s were still good for two special but secretive birds: wood warblers and hawfinches. They divided the year for me. Wood warblers were spring and summer visitors, and hawfinches were mostly autumn and winter birds. If I saw either bird the day was made. Once I had found them and knew them to be there, my entrance into their places—the final stretch of my delivery round, perhaps three hundred yards of road at the wood edge—was always charged.

Both birds were hard to see. In a secondhand book shop I had found a pamphlet published in 1907, *Some Common Birds of the Neighbourhood of Clifton* by Herbert C. Playne, that described the wood warbler as "extremely abundant" in Leigh Woods, but they certainly weren't by the 1970s. There were a few; they were there for just a few weeks from the late spring, and they lived high and hidden in the top branches of the newly green trees. I found them only when the males sang and made their parachute song flights through the leaves that shuffled and flickered with the same color and movement as they did in the breeze. After that they disappeared silently into the canopy, like just another leaf, before their invisible departure later in the summer back to Africa.

Hawfinches were even more elusive. I saw them fifteen times in five hundred delivery rounds. They were very shy birds, the color of autumn leaves and ripe hazelnuts. I knew them by their exits, their short wings and short tail flying away quickly through the trees, rebuking me for having broken into their arcadia, for catching them at their bath in a woodland puddle like Diana, and once trying the tarmac of the road, stepping like firewalkers over hot coals.

Hawfinches have a voice as near to silent as any bird. Their call is a tightly squeezed *pissp*. It is very easy to miss. The birds seem to be in a permanent reverse out of the world. But for their beak you would think there was something apologetic about them, with their gentle tones of camouflaging green and brown, their whispered call, and their swift retreat from view. Their beak, though, is enormous, the largest and most powerful beak of all the European finches. It is a colossal tool, an anvil and hammer at once, and very serious. You might call it out of proportion except that nothing is in nature. The books told me that a hawfinch could exert enough pressure with its beak to squeeze a cherry stone until it popped and surrendered its soft heart. Their call makes this very sound.

At best I got fleeting glimpses, but both birds had great power for me. On an ordinary weekday afternoon with my bag full of local news, just a few seconds with them opened another world. They were calm but wild. On the back road I felt that only I knew that the wood warblers and hawfinches were there and that their quietest of dramas, whispering themselves out of existence, were shown only to me. Sure enough, they have vanished now. I haven't heard wood warblers in Leigh Woods for twenty springs, and the hawfinches have gone too.

Even then, there was a bird missing. The gorge needed another, different performer. As I crossed it every day I always thought it should have peregrines. All the extra air the gorge put into the world seemed made for them, and them for it. A stooping peregrine would fully chase it into life. I looked and hoped and craved a stoop, and made one in my mind and replayed it again and again: that hinge in time when the peregrine reaches its tipping point and the accumulated knowledge and energy of the bird's hunting life—its eye on its prey and its height above it—rolls it from effortless horizontal flight into a near vertical dive. It stops in the air, appears to raise itself slightly, folds it wings over its back, and, from the most casual of starts, slips forward. It drops in the shape of a smooth pebble down, fast, faster, falling in an ecstatic arc toward something far below it that the falcon has seen and singled out but which will probably never know or see what is about to hit it. It is a piece of site-specific theater, a superb stunt that is ordinary but beyond our imagination, a free-fall made by a controlling intelligence beyond our comprehension.

One November afternoon slicked with rain, I was looking out for peregrines as I cycled to Leigh Woods, my newspaper bag over my shoulder. The day was quiet. Bikes had to go on the pavement of the bridge. Walking about twenty yards ahead of me was a youngish man with thick, tumbling black hair. He wore a dark brown corduroy jacket. I began to catch up with him and as I braked, preparing to go around him, he turned his head to the right and glanced over his shoulder. Perhaps he saw me, maybe he didn't. As he turned his head he brought a cigarette in his right hand to his mouth and dragged on it; its tiny fire throbbed brightly back to me. Then, still walking, with the cigarette in his mouth, he put both hands on the wooden railing of the bridge to his left and vaulted over the side.

I knew at once what was happening and what had happened. In front of me, a man was jumping off the suspension bridge to kill himself. I stopped my bike, dropped it beneath me, knowing that I needed to hurry. I didn't look over the railing; I couldn't. It was as if a steel wall had grown up from it high into the sky, stopping me from looking down after him.

I turned and ran back to the toll booth about one hundred and twenty yards behind me. I leaned through the glass slits into the booth, where pedestrians passed their 2p fare, and said to the man inside, "Someone has jumped over the bridge." I was hurrying, because in my head I thought it might be possible to stop him, to catch him even.

There were two men inside the booth; they both pointed to its side door and I came round into their little room, with its tea mugs and transistor radio and low piles of copper coins waiting to be given as change. One of them was already on the telephone. I remember him saying, "We've got one over." His voice was gray like old paint. I was breathing hard. The men were calm. One asked me if I'd left my bike on the bridge and I said that I had, and he said he would come with me and get it. We went out across the bridge. The steel walls were still up for me, but as we got to where my bike lay, he put both his hands onto the varnished railing, just as the man before had, and peered out, over and down to the river and the road, two hundred and fifty feet below. I couldn't look down, but looking back toward the end of the gorge where the river arrives from Bristol and the high ground gives way to ordinary England, I could see that far below there was a line of cars on the road that ran beneath the bridge.

We went back to the toll booth, me pushing my bike, the bag of papers still around my neck. I couldn't form words in my mouth without swallowing hard. The men didn't speak to me; they finished their tea. An ambulance arrived outside the booth. I thought it might be for me, but one of the men went out to explain that it had come to the wrong place. The newsagent appeared—the men must have telephoned him—and he took my bag and drove off over the bridge to deliver the papers. It grew dark. In the booth, I waited. A policeman appeared and came in, big and clumsy with his helmet. He asked me what I had seen and my name and address, and wrote a few words in his notebook. He said he would come and see me again and that I should go home.

Apart from glimpsing the line of cars backed up on the Portway, I had no sense of the end of the man's jump. All I had was the beginning. As I cycled home a silent film of his leap began to play in my mind. It ran again. Then again. I told my parents what I had seen. They were kind, but I could see them struggling to know what to say or do. That night was terrible. I needed to clean my mind and stop the film and I couldn't. Over and over, that leap and the fall. The leap and the fall. The leap and the fall. I slept in my parents' room on a camp bed.

The next day I had to do my paper route as usual. Someone was walking ahead of me on the bridge again. I got off my bike and waited for him to get to the other side so that I could cross. There were jackdaws, but I couldn't look down over the railings at them as I had done before. Since then I never have been able to.

As I delivered my papers, I became preoccupied with thinking that if I had been there a minute earlier or later, I wouldn't have seen anything. This was a thought-game I was used to playing because of birds. The hawfinches on the road were only there for a few seconds. If I hadn't been there then, I would never have seen them.

The second night was worse. I couldn't close my eyes without the half-turned head, the glowing cigarette, and the athletic vault flashing there again and again, stuck in the gate of my head. I felt I'd been got at, contaminated somehow. The world was calling on me. Life and death, already. Everything felt sped up.

A few days later a policeman and policewoman came round to our house and asked me to describe what I had seen. It didn't take long. They wrote it down. The policewoman told me that there would probably have to be an inquest into the cause of the man's death and that I would have to give evidence. No one said the word "suicide"; no one said anything about what would make a man jump off a high bridge. It was as if I had witnessed an accident, not a decision.

As I waited for the inquest, I began to extend the man's fall in my mind. I think this was because I wanted to get the endless replay of its start out of it. But I think it had to do with birds too. His launch, and then what? There was a moment of energetic push, out into the air, powered by the kick of his legs off the bridge and his hands on the railing, and then maybe he registered the thinness, the nothing, as the air failed to hold him. But my man—he became "my" within a day or so, as I grew aware that I was carrying something of him with me whether I liked it or not—my man didn't want to stay up.

My mind closed in on the moment when the energy of his push from the railings equaled the force of gravity and the air's collusion with it and his body felt weightless in space. There would have been a moment like that—that was what it felt like when I used to jump from the swing in our old back garden. I loved that feeling. For a fraction of a second you have the tiniest intimation of flight.

A peregrine could have shown me this if I had seen one begin its stoop. A man did instead. The peregrine uses its own falling weight to bring death to its pigeon prey. My man used the same to kill himself. It didn't matter that he couldn't fly. It mattered that he didn't have wings. There was nothing to hold him up. The peregrine folds it wings away so the same thing happens, falling and killing. The man fell and died.

Weeks later, a policeman called by at the house to deliver a notice about the inquest into the man's death. I had to attend the coroner's court in Bristol. It was on a school day, in the morning. I wore my school uniform and my father took me. We arrived without being certain of where to go. We went into the building, its municipal wooden doors with crinkly glass and polished brass plates giving onto a stone hall with a desk and rooms off it. We were directed to a room; it was carpeted and half paneled with wood. The stone busyness of the hall was muffled and everyone in the room was whispering. There were only a few people there. One was a youngish woman, her hair tied up, wearing a black dress. Hers eyes were rimmed with red as if a razor blade had shaved a single layer of skin in a fine circle, taking her eyelashes away to show what was beneath. We waited in the quiet.

We were ushered into the court. There was wood everywhere. The floor was wood, there was paneling up the walls, the court furniture and its strange low fences and gates, which had to enact some separation of parts of the court—all of it was wood. I can't remember the order of events but quickly, it seems, I was called up to the witness stand. I think I had to swear on the Bible that I would tell the truth. An official held a card in front of me with the oath that I had to say. My throat was dry and I tried to lick my lips discreetly to help me get the words out. The coroner in a dark suit sat like a judge behind a wooden desk on a dais. The only windows were behind me, set high in the wall. A functional, neutral light came through them, as if the city had been able to order a civic sky.

The hush was like a library or a museum. I was on show, standing up, and expected to speak loudly. I knew that amongst the people sitting below me in the public area of the room would be—surely—someone who knew the man. The youngish woman with the eyes was there.

The coroner asked me to say what I had seen on the bridge. I was allowed to read from the statement the police had taken. It was only a few sentences long but it was hard going. I never seemed to have enough breath. Each sentence fell away and was lost in the hush and the wood of the court.

I got to the end of my statement. The coroner said, I remember, "That wasn't very impressive, was it?" and he looked across at me. He said the court needed to be able to hear me clearly and he made me read it out again. What I wanted to say was that I was frightened of what I had seen on the bridge and that I hated the man for doing it in front of me. But I felt under some official pressure to send this dead man out of our world and that the job had fallen to me. I tried again with the words. The wood of the room seemed to soak them up once more. I got to the end of the statement. I was a disappointment, I knew, and I hadn't been able to do what I should have done. I could tell that the coroner was still not impressed but had decided he had got from me as much as he could. I was allowed to sit down.

The court went on. There was another witness. A middle-aged lady had been driving along the road under the bridge. Something had hit the rock face to her right and bounced onto the road in front of her. It was him.

Someone, perhaps the coroner himself, told the court about who the man was. I don't remember his name. He had parents who lived somewhere in the Caribbean; they hadn't come to the court. He had been depressed for some time. His business hadn't gone well. He had run a shop on his own, just up from the suspension bridge in Clifton. As this was described, I remembered walking past the shop front one day and catching his eye as he looked up from behind his counter. He had had a mustache.

Ten years after my falling man, when I was collecting information on the threatened birds of Africa—a catalogue of rarities and extinctions substantially made from information taken from the labels attached to dead birds' feet—I was struck by a sense of something familiar in the atmosphere of the Natural History Museum's bird collection. As I went into the building at Tring, just north of London, for the first time, my breath faltered. The drawers of the dead birds that I opened, the wooden and metal cabinets and the hospital-clean corridors between them, the place's institutional quietness and the feeling it gave of being in some giant coffin—I had been somewhere else like this. And the thought rose up in me, drying my throat, that the dead birds of the museum were taking me back to the coroner's courtroom in Bristol.

After fifty years of absence, peregrines came back to the gorge. Now they are easy to see there. The watchers sitting in their deckchairs on the cliff top will point them out to you. I have often seen them in the gorge and over the city, but I still have never seen one stoop.

December

Black Birds and Black Nights

An Airy Crowd came rushing where he stood:
Which fill'd the Margin of the fatal Flood.
Husbands and Wives, Boys and unmarry'd Maids;
And mighty Heroes more Majestick Shades.
And Youths, intomb'd before their Fathers Eyes,
With hollow Groans, and Shrieks, and feeble Cries:
Thick as the Leaves in Autumn strow the Woods:
Or Fowls, by Winter forc'd, forsake the Floods,
And wing their hasty flight to happier Lands:
Such, and so thick, the shiv'ring Army stands:
And press for passage with extended hands.

VIRGIL, TRANSLATED BY JOHN DRYDEN

. . . as we take, in fact, a general view of the wonderful stream of our consciousness, what strikes us first is the different pace of its parts. Like a bird's life, it seems to be made of an alternation of flights and perchings.

WILLIAM JAMES

Dusk on the winter solstice: the shortest day and longest night of the year. I am cold and alone on a track on a peat moor, looking toward the dying light in the west. Stippled across the sky in front of me, like the breath of the earth, are thousands of starlings arriving to roost. They are going to put away their day and so, too, the year.

Tomorrow the glorious creep toward spring will be under way: more light; a future; repairs; song, nests, and eggs.

The year has drained to this day and its paltry hours of watery sun. In the middle of the afternoon, a cold, iron-hard dark arrived from the east and pushed all the light away to a buckling golden foil fussing on the western horizon. There the day launched a last flare, like a crack of magma seamed through lava. The year was burning down.

It is freezing on the old track across this peat moor in Somerset, in the west of England. Last night's unmelted ice is thickening, reaching toward tonight's dewfall. It frosts the tiny flashes of the puddles, the lank grass around them, and my boots. Molehills along the track have frozen mountain-hard with little ice caps on their summits. The sky is clear and harsh. High ice crystals prink its blue like a snowfield. Night is coming and I am shivering, but from all points of the evening skies the starlings come over me like warmth.

Westhay is part of the watery grid of the Somerset Levels, which are level or flat relative only to the hills that rim them on three sides. Their flatness is like the flatness of the sea out to their western edge. The land buckles as if it remembers it came from this sea. Around ten thousand years ago, at the end of the last ice age, when the sea rose Westhay became a salt marsh. Six and a half thousand years ago, as the sea receded, the marsh silted up with a rich tilth and became, in slow succession, a freshwater reed swamp, then a wet fen woodland, and then a raised bog.

Millennia of rot made the peat that cloaked these new flatlands with a deep black soil. More than five thousand years ago people built the first trackways across the spongy soak, like the one I am standing on, so they could pass to and from adjacent higher ground, the hills of the Mendips and the humps and mounds of Glastonbury. At Shapwick, near Westhay, in the spring of 3806 BC, the Sweet Track—a wooden road—was laid, using cut oak, ash, lime, hazel, alder, and holly. It is amongst the earliest piece of woodmanship in the world. Later, people began to dig the peat. Walking on the track I can feel the earth's plastic bounce, its five-thousand-year-old give.

Nowadays something of an old, low-intensity life lingers on the levels in the slow suck and churn of mud and grass, in the drip and spread of water and sunlight. I can see where an abandoned orchard's unpicked cider apples ferment beneath the skeletons of their trees, red globes of bright treasure in a damp grave. Bullocks chew an endless cud in the fields. Their hot breath hangs at their nostrils as they breathe out their own local weather systems in the twilight, making clouds from a cast of their lungs.

People, starlings, cattle, and peat cutting all lived side by side until the recent past. In the last few years the commercial digging stopped, and the gouges cut into the land were abandoned and allowed to fill with water and new vegetation. At Westhay the old peat workings have left flooded mires and swampy pools with reeds growing across them, which make a perfect nighttime sanctuary for roosting birds.

Every evening, through the winters of the past few years, thousands, even millions, of starlings have come to sleep here. Eight million were counted, somehow, in one roost here a year or so ago. This may be the largest-ever gathering of birds in Britain. Imagine Hyde Park Corner or Central Park in New York City in failing light and the entire population of London or Manhattan arriving there from all points across the city.

At Westhay some starlings are local birds, others fly in from miles away. Arrow-headed echelons of them had shot south above me as I'd driven to the levels hours before. Individual starlings are sharp and pointed, and the first flocks they form on their way to their roosts are sharp and pointed, too. They were already heading purposefully toward Westhay, raking the sky. In midwinter the starling's day is even shorter than the few hours of light. Not long after midday they are thinking of bed, and their assembly flights begin.

Other starlings spend their winter days in the fields adjacent to the reed beds where they will roost, and there I find the first gathering birds. Flocks of a hundred or more already are out on the levels. Local birds and arriving birds mix together, squabbling, feeding, and talking. In the fields they look older and more purposeful than they did as village starlings an hour before. They seem to be joining some necessary action. A call-up is under way.

This evening, summer is farther off than it will ever be. Stowed sunshine from months ago is being rationed, like the last grains of sugar in a siege. Its light and heat survive in only the flimsiest of things: the feathered seed heads of the reeds that engrave fine scratches onto the plate of the sky; the tiny contact calls swapped between parties of long-tailed tits as they move through the willow tops, living in the warmth of their own talk. Everything else is, or soon will be, a shade of black.

The light has nearly gone. All that is moving looks shadowed: the great spotted woodpecker whose bouncing flight ahead of me echoes the folds and dips of the fen path appears to have lost its white and red; the buzzard that planed alongside the ditch looked so dark that I thought at first it was an aberrantly dark melanistic bird; the gulls that beat north look no whiter than the cormorants going the other way.

Black has overtaken the starlings, too. If I could separate just one bird from the lines flying above me, or the legions in the fields, I could find their daytime sheen of pearl-spotted oily iridescence, but the massing birds take on a generic black. It is not the furred black of wet peat; that will come later. For now the starlings' black is the feathered brown-black of drying peat, the soil at the surface, not the buried earth.

From all sides there are lines of starlings, in layers of about fifteen birds thick stretching for three miles back into the sky and coming toward the reed beds that surround me. They come out of the farthest reaches of the air, materializing into it from far beyond where my eyes or binoculars can reach in the murk. All fly with a smooth, lightly rippling glide, as if the net that they are making of themselves is being evenly drawn in to a single point in the reed bed.

Their arrival and accumulation has been eerily silent. From the early afternoon, first in the villages and then in the staging fields, they had made a great noise, a collective telling and retelling of star-

ling life that rose through that hour of pre-roost talk to a complicated but loquacious rendering of all things: idiomatic adventure,
mimetic brilliance, and delighted conversational murmur. Once this
annotation of the day was done, the birds grew quiet and lifted up
and off to begin their condensing flights in toward the roost.

Thousands of mute birds are all around me now, their wheeze
and jabber left behind. Many thousands more are too far away to hear
but their calm progress toward the roost suggests they fly in silence.
Closer, the only noise is of the flock's feathers. As they wheel and
gyre en masse, the sound of their wings turning sweeps upward like
brushes dashed across a snare drum or a Spanish fan being flicked
open, making a brittle percussion on the skin of the sky.

The air is thick with birds, inches apart and racked back into the
darkening sky for a mile. Every bird is within a wing stretch of another. None touches.

A rougher magic overtakes them as they arrive above the reeds.
Conjured balls of starlings roll out and upward, shoaling from their
descending lines, thickening and pulling in on themselves, a black
bloom burst from the seedbed of birds. Great cartwheels of them
are unleashed across the sky. One wheel hits another and the carou-
sels of birds chime and merge. They are like iron filings made to
bend to a magnet.

The flock—but "flock" doesn't say anything like enough—
pulses in and out. My eyes are forced to deepen their field of view to
take in birds behind birds behind birds; my brain slows as it tries to
compute the organizing genius of what is in front of me. Floating
above a peat bog in the dark, I am taking an exam in arithmetic in a
room drawn by Escher. I cannot do it. Optical illusions make me feel
sick. To describe the flock is like trying to hold on to a dream in daylight—it slips from me; it cannot be summoned, except in fragments, and those cannot be transcribed. Try singing it.

I think of Thomas Tallis's forty-part devotional motet "Spem in
alium" (from 1570). For as long as I have known it I have loved it,
especially in the recording made in 1962 in King's College Chapel in

Cambridge. Could its eleven and a half minutes of singing light the black midwinter night and the black midwinter starlings?

"Spem in alium" doesn't describe what the flock does. It is the flock. The music—unaccompanied singing, or rather singing that only accompanies itself—comes in, opening its throat before us, beginning with some tentative note on some frontier of sound, arriving into a space from a place without space, from far away. It might be one bird flying or the sound of a wave beginning far out in the Atlantic. The sound catches and swirls toward us, becomes a striving, and folds into itself and floats and opens further with a beautiful frail young solo that twists my ears and then gives onto a landscape like the great slab of an abstract painting, a Peter Lanyon sky masterpiece, starlings hatching from the evening. It is huge and everywhere, but tilted and very close. And all along there is the strangest of pulses, a breathing, a flexing continuo, that rises into the heights of the chapel at King's College, climbing like a filling bath around the stone. The voices arrive from all corners, unfolding and bending, in relay and alone. Forty throats open. They sing together and against one another, and then against one another and together again. It is a wind blowing out of paradise. It is a vast river of warm stone and dark skies, of sea silver, of black and sheen and matte and dust and lisping and echoes and news and pain, and it deepens beyond voices until the great stone room is singing its own song, and its sound goes brightly down beneath the building into the earth and then rebounds, coursing up the vaulting out into the winter night. There are snake whispers and dead leaves rustle; there is music for outriders and prophets, songs for latecomers and dreamers. It is a coffin and a bed. Then: a resolution, the first of many sung "Domine" that wrestles with the flex of moving song, which wants to gather in the word right away and take it home but cannot; the word repeats and rises and wanders beautifully out into terraced voices, with vistas that stretch as far as the mind can go, pushing back the night, opening the earth, and then the lit dark comes, its stars thrown over you the moment the sun goes, and the soil sings. And big and kind and at the right time it ends, calming to begin with, an embrace and a decision, and then the wild shriek of a single young voice climbing and breaking back out into the sky for one last flight, loud, and screaming, and elated.

Standing under the storm of starlings, I wonder about the moment when a single bird commits itself to the collective swirl. You can see it. Individuals hurry toward the flock; one comes quick and low just over my head from behind and surfs into the fabulous mêlée. I try to keep it in view and for a few moments I have it. Then its singleness is taken into the whole as it angles its wings as all the others do around it and the scattering shot of birds shifts and I lose it.

We cannot watch without thinking that the flock becomes a single organism, but why should it? There doesn't need to be one god up there; there could be eight million.

I can feel the starlings wanting to go down into the reeds, toward the safety of the soil. Each flick pushes them lower but still they rise and fold upon themselves. Their whiplash just above the reeds sets off water rails. Eight siren squeals start up across the marsh, as if the pressure snap of the starling flock above has forced open a door so that these spook sounds (of a torturer or the tortured, I am never sure) might escape from below.

It is almost dark now. It is getting colder still. Freezing mist rises from the gelid pools. Wigeon whistle like boatmen hurrying to pack away their oars. Pied wagtails *chissick* about me on the track, loath to settle before the starlings.

Now they begin to disappear. The lowest birds brush down over the reeds again. Half continue, but under the flowing cape of the larger flock, a phalanx has invisibly dropped down. Starling fission. Why those birds, I wonder, why there? Have they been pushed to the ground by the bulk of others above them, have they finished something those birds are still occupied with? Has this been a purposeful dance, are they leaders or losers, are these marked as their reeds, is this their dormitory? Starlings' roosts are organized with adult males predominating at the center and young females pushed to the periphery, but how is each bird filed into place?

There is a rumpus. The birds that slipped from flight down between the reed stems at the edge of an iced mire have disturbed a bittern. "Myre dromble" was an old name for a bittern, meaning a sluggish bird of the marshes. It gets up like a sleepwalker and flaps heavily away through the flying starlings, an old unrolling brown carpet, until it is able to slump down in another patch of reeds.

The combing and pleating of the master starling flock continues. But the exit is quick and simple. The gathering preamble, the flying and flocking, has been the main event. Deciding to go to bed takes much more time than getting into it. More and more starlings are dropping into the reeds, all immediately hidden from view.

As they arrive at their roost they begin talking again. The encore takes one minute only. They share their last words and then a simultaneous full stop. The water rails quit squealing the moment the starlings stop their noise. Shush. The loudspeakers are switched off at the same time as the light. The show is over; everything has been put away into the black box of the earth. Good night.

I stay on the track in the midwinter dark looking across the reeds in front of me that held thousands of invisible sleeping starlings. I remember the storm petrels on Mousa waking at midnight in midsummer, the black tower and the black birds that came from the sea to it. Now at its other corner, the year is ready to shut. Behind the reeds, the last vestige of light seems to keep an auroral trace of the starlings. As I look, for a few moments, there is a kind of black starling light twinkling there. For a few moments there is a kind of black starling light, like winter pollen, twinkling amongst the starlight, an after-image as starlight is itself, of what was once there.

I walk back down the track and am surprised to see shadowy figures joining the main path from other tracks across the moor. Unnoticed by me as I stared at the starlings, others had come to see the spectacle. A scattered crowd watching an assembling crowd come down to earth among them.

As I drive back in the dark, they seem to be everywhere still. I think of the starlings I saw on my way out gathering in bare winter trees in the villages of East and West Harptree and making their noise, and of Coleridge, who also watched starlings, writing "The Eolian Harp" just a few miles west. And I remember that Cecil Sharp in 1904 had collected bird "starving," or scaring, cries there too: "Hi! Shoo all o' the birds / Shoo aller birds / Shoo aller birds."

Back at home, I want to create my own starling diary before bed as they do. I pull books from the shelves and scrabble in notebooks and plunder my memories, and while the Westhay birds sleep, I gather and write down as many of my starlings as I can:

The starlings running over the snow-covered lawn of our back garden, doing their human-like walk but printing their bird feet in the snow, one of the first birds I identified at age five or so;

Starlings weaving through flocks of parrots in Los Angeles, the Europeans jiving with the Brazilians;

The starlings of Fair Isle that go south in their first winter, return home, then never leave;

Nero's starling, as told by Pliny, practicing its Greek and Latin and managing "ever longer phrases" every day;

Starlings in the rain on the back of sheep in Scotland, like people riding elephants;

Bobby Tulloch's scientific note on starlings from Shetland that use sheep as towels, and another of a starling drowning there in a bowl of custard;

Rose-colored and regular starlings in a thistle field at the foot of a Turkish mountain, children in starling pink and starling green watching me there;

A starling singing its song of songs including an oystercatcher's yelp and a curlew's summer bubbling on the roof of the house where Van Gogh lived in south London, telling me where it had been that summer;

Starlings at the railway station in Bristol baffling a predatory sparrowhawk by opening and then closing their roost flock around it;

Introduced starlings, the first birds I saw in South Africa, when being met at the airport by Claire, who grew up in Cape Town calling them Eurostars;

Bertolt Brecht and Rosa Luxemburg identifying starlings with the urban proletariat;

Starlings in Dante compared to the damned;

Edward Thomas in France in March 1917 describing German shelling "over our heads into Beaurains all night—like starlings returning 20 or 30 a minute";

Starlings flying across Columbus Circle in New York City in 2008 toward Central Park, where they were first released in 1890–91;

Rachel Carson standing up for the maligned introduced starlings of North America—which only did what other Europeans had done: arrived, moved west, succeeded—writing an article in 1939 called "How About Citizenship Papers for the Starling";

An English roost of starlings detected on a radar screen in the Second World War, disturbed by German V-1 bombs flying over it and "causing a scare that the enemy had invented a new form of radio jamming";

Starlings blamed for £100 of damage to a reed bed in Lincolnshire in the eighteenth century, after one night of a roost and its "abundant excrement";

Spotless starlings laughing at me, so it seemed, from the roof of a holiday apartment, near Alicante in Spain, as I tried to learn to dive, aged forty, in a swimming pool below them;

Laurence Sterne's caged starling in *A Sentimental Journey Through France and Italy* that speaks its confinement: "I can't get out—I can't get out"; Sterne's own name, close to "starn," a dialect word for starlings;

The starlings that bred last summer opposite my apartment, in a roof crack above Creation Nail Studios, and wheezed from the gable through the rain and traffic slush to my desk;

"Starlings!" shouted derisively to the men on the quay, by passengers on ships from the northern isles of Orkney as they steamed into Kirkwall;

Coleridge keeping starlings in mind, in a letter from him explaining why he wasn't writing, remembering Sterne (as well as Macbeth) and comparing himself to a starling: "self-incaged, & always in the Moult, & my whole Note is, Tomorrow, & tomorrow, & tomorrow";

Coleridge, two years before that, watching winter starlings from a coach traveling south to London and writing starlings as well as Tallis sings them: "I saw Starlings in vast Flights, borne along like smoke, mist—like a body unindued with voluntary Power—now it shaped itself into a circular area, inclined—now they formed a Square—now a Globe—now from complete orb into an Eclipse—then oblongated into a Balloon with the Car suspended, now a concave Semicircle; still expanding or contracting, thinning or condensing, now glimmering and shivering, now thickening, deepening, blackening!"; and two days later, in the turmoil of London, Coleridge contrasting "the harmonious System of Motions, in the country & every where in Nature . . . [with] The immoveableness of all Things thro' which so many men were moving";

Starlings dropping down to my sister's lawn seconds after she has scraped the crumbs from a bread board onto it;

Starlings hovering at a seed dispenser in my parents' garden, irked but cunning, trying to work out how the tits manage to feed there;

A young starling with its legs splayed like a baby in diapers going headfirst down the stretched throat of a lesser black-backed gull near my home in Bristol;

A shambling family of starlings on Coll in May, debauched and chaotic and moaning about winter, above a corncrake still stoking up the year with its *crex crex* song ("endorsing summer" as Louis MacNeice says, with the "sourdine in their throat" as Andrew Marvell says);

The "brabbling" starling, as the poet John Skelton describes its bubbling song;

The charivari of morning starlings on countless roofs of my life, jangling and wheezing and putting the kettle on, always installed in the day long before me;

Fir trees full of starlings, common grackles, and American crows in Fergus, Ontario, defragging the day before going to sleep;

Starlings perched but swimming with their wings as they sing their sex songs on a farm near Bordeaux, mimicking a tawny owl's *kewick* (did they dream it?), a golden oriole's torrid whistles, a green woodpecker's *yaffle;*

W. B. Yeats's "stare," and John Clare's "starnals" that "darken down the sky," and him in his lunatic asylum thinking he is in prison: "I am still wanting like Sternes Prisoners Starling to 'get out' but cant find the Way";

And my childhood sleeping under starlings, listening to them dragging junk for their nests into the cracks of the roof above my bed, their scuttling noise up there at dusk and dawn;

My starlings and yours, always going home, all of them, at home everywhere, all of them, running with the grain of the world.

I started writing the beginnings of this book in an attic room in my parents' house in south Bristol, having fled there sad and distraught in the middle of a winter. My domestic and emotional life was a mess. Trying not to feel sorry for myself, I sat at a table under a window and began to write about what I remembered of birds and my bird-watching forty years before, when I had lived with my parents as a boy. As then, gulls trekked into the city along the lines of water that flowed through it, like flakes of gold leaf against the evening sky. In the morning they went the other way, their marine bleats tugging Bristol out to sea. From the high window I looked north beyond the gulls and back toward the city and the steep bowed hillside of Clifton with its Georgian terraces like sculpted continuations of the gorge's limestone cliffs.

I had lived in another attic in a house up there as a child. Starlings scratched at their nests in the roof an inch away through the plaster behind my bed; below me, I had watched spotted flycatchers in July breeding in the garden wall, coming to and from a crack between the stones where the mortar had loosened and crumbled to dust. The snap of their beaks was the only sharp noise in the hum of hot summers. There too, one May, I had identified a garden warbler singing hidden within a huge candled horse chestnut that spread its great branches across the road as if it were holding its own skirt. For three springs this bird or its children sang from this tree. I never saw them once, but their song filled my little room, a rich, convoluted slurp of a song, which still sounds to me like a commentary on the color of their life, "a green thought in a green shade."

Forty years on I repaired myself as best I could and, after a few months in the new attic, found someplace of my own to live. A year later my parents also moved and I came back to the house to help them clear the loft space of years of family clutter that had been shipped from house to house in Bristol. Among my old school books and some Victorian stuffed birds, which I rescued and have around me now, there was also a box of fifty small plastic bags of soil. Some were peaty black, like dried moleskins; others had a more southern Somerset ruddiness, a color that to me always seems old, like cheese rind or the face of a red deer. Each bag had a label stuck to it, with scrawny writing listing a grid reference and sometimes a field name, but many labels had lost their glue, slipped off, and gathered in the bottom of the box.

Thirty years before, I had collected and bagged the soil as part of a barely begun and never finished school geography project on the land use in the Somerset Levels. Only when I saw the bags did I remember having done this. I had dug all that soil aged seventeen with my mother's garden trowel from fifty fields between Westhay and Wells, intending to map its qualities and correlate this with land use. Did cattle need richer soil than apples? How moist a loam could vegetables tolerate? It was a ludicrously ambitious project and I ended up doing two days of sample collection and nothing more. I don't know why I had held on to the soil. After it slipped my mind it had found its way up under the roof. The whole house had been living under the soil of the Somerset Levels for years. And I had come back there seeking refuge, a roost, and had slept and written in an attic room beneath the earth: black thoughts under black soil.

In the Campo Santo in Pisa, the city's beautiful colonnaded burial ground, I once watched a house sparrow that had been picking at the earth in the central garden of the building, as it flew in through the open tracery of a stone window, along an arcade of graves and tombstones, and out through another window. The Campo Santo was founded at the end of the twelfth century when Archbishop Lanfranchi returned from the Crusades with a cargo of soil from Golgotha in order that eminent Pisans might be buried in holy earth. The sparrow moving through the building made me think of Bede and his image of life's journey as a sparrow flying through a lit hall. The echo the stone gave to the sparrow's chirps as it flew made me think of Nick Drake's 1969 home recording of his song "Fly" (released on *Time of No Reply*). Before he begins to sing his song—which is about discovering that the earth is where we belong—house sparrows chatter an accidental chorus outside the room where he made his tape. Against the sound of sparrows in the gutters, he sings, "It's really too hard for to fly." And I thought of the sparrow above the dead of Pisa, and Nick Drake dead now, and all the sparrows that sang from his gutter dead too, and again Keats's line from his letter of November 22, 1817, to Benjamin Bailey came to me, a talisman ever since I first read it years ago: "If a Sparrow come before my Window, I take part in its existence and pick about the Gravel." Keats's sparrow is dead. Yet every time I read his words, as every time I hear Nick Drake and his sparrows or hear the singing made in King's College Chapel in 1962, I am restored to life, as I was in the dark of midwinter, at the year's midnight, on that moor amid starlings.

January

South

You acid-blue metallic bird,
You thick bird with a strong crest,
Who are you?
Whose boss are you, with all your bully way?
You copper-sulphate blue bird!

D. H. LAWRENCE

What birds were they? He thought that they must be swallows who had come back from the south. Then he was to go away for they were birds ever going and coming, building ever an unlasting home under the eaves of men's houses and ever leaving the homes they had built to wander.

JAMES JOYCE

The new year begins with dead things, with blown eggs and dried skins, on a tobacco farm in southern Zambia where an old British army major, poleaxed with illness, deplores the end of all good genealogies amid his cabinets—the best collection in the world—of eggs and skins of cuckoos and other avian cheats.

The year starts with me feeling disinherited and estranged. I am, at best, third in line to a throne. Farmhand Lazaro is first: he has found a tawny-flanked prinia's nest in the morning, before his work in the major's tobacco crop, and leads us through the tall wet grass of an abandoned field toward a marker flag of red tape. Claire, an ornithologist, follows him, in her hat and long sleeves, guarded against the sun and shouldering her bag of equipment. The farm is her field site; the prinias, cisticolas, and cuckoo finches that parasitize them are her study birds. I have come to see her and them and bring up the rear, guessing at what I am seeing, sweating and lost in Africa.

Then a party of four swallows flashes past, drawn perhaps by the insects we stirred. They dip near the grass heads between Lazaro and Claire and fly on, their dark blue backs catching my eye. Something shifts and I can see again. They are barn swallows, my swallows.

The night before I left Britain to visit Zambia I had been in bed in Bristol, surrounded by my half-packed bags; I hadn't shut the curtains as a big near-full moon was riding the night sky. Like a bedridden astronomer, I lay in the dark looking at it. As it barged through the thinning clouds I picked up my binoculars. A cold ash-gray winter moon with none of its yeasty summer curd, it looked like a hole cut in the sky's black curtain. In my childhood fantasy, it was just that, an escape hatch, a way through and out. It itself was nothing, just a hole, the mouth of the cave we all lived in, from which a stone had been rolled. Through the hole you could see a circle of the world that lay at the end of the deep space far beyond our space. I wanted to go there.

As I watched through binoculars, three birds crossed the moon's light, silhouetted, turning toward it as if to go through its hole. I lost them. They were probably redwings; the sky of Bristol is pricked with their thin wiry calls most winter nights, as they move into mild southern England away from the frosted hardships of Iceland or Scandinavia. The invisible sun's light reflected off the moon, making backlit shadow puppets of the three birds, picking them out from an otherwise hidden flock, all carrying on their migratory business in the dark. The moon was once thought to be the destination of migratory birds like these thrushes. Watching for a moment, the redwings crossing this lunar stage, it was possible to understand where such ideas came from.

The journeys the swallows had made from northern Europe to fly above us in Zambia seemed as incredible as the prospect of a red-wing flying through the moon's hole toward the silvered light of a world beyond. When our summer birds leave us in the autumn it is very hard to follow them in any way, to imagine their journeys south, their wintering quarters, their new neighbors, the way they become African. We let them go and take up with newcomers from the north, such as redwings, who have arrived to fill the spaces left by the departed. The birds in front of us become the truest, most arresting things. A swallow in midwinter would be a mistake, a lost or sick bird.

But in Zambia there are plenty of swallows in January, all of them at home. The four above Lazaro and Claire are part of the fine black mesh of swallows, martins, and swifts, half European wintering birds, the others local migrants or resident species, all of them loosely stitched through millions of acres of African sky. Seeing a swallow above the long grass of a Zambian farm brings on a kind of happy vertigo in me. I might have seen the same bird above the reed bed at Chew Valley Lake near my home in Bristol last summer, might even have ringed it there, held and measured and released it back to its sky there.

Before I saw the swallows I was struggling to see anything. Flying south from London I had known that all of Europe's summer birds were spread out below me through Africa. This should have been reassuring but I felt disconcerted. As we crossed the Mediterranean, five miles up, I peered out my porthole and looked down at the beginnings of the Sahara, the citrus-peel soil of Tunisia and Libya giving way to an antelope skin, pulled taut over the bony skeleton of Niger and Chad. There were fields that from the air looked like blank pages dropped from a book, and each dotted grid of an orchard looked like an uncountable abacus.

I wanted a bird in the bush, a legible mark on the earth like "the three orange trees at the edge of a field near Guadix," that Saint-Exupéry writes about, guiding flyers through Spain. But all I had was an idea of unknown numbers in an unreadable place. Down there were all my redstarts, but the landing lights of their quivering red tails were snuffed out by the brightness of the day and the vastness of the place.

After dry fields and trees came the sand, shadeless burning miles of it, with long-running ridges of rock making scarified marks on its otherwise featureless face. It takes a willow warbler between twenty-nine and forty-four hours of continuous flight to cross the desert. Some do it without stopping; others drop to the sand in the heat of the day and rest, flying on in the cooler dark. I thought of the sedge warblers I had trapped at Chew and banded, blowing apart their breast feathers to assess their fat reserves. The tiny sachets of it laid along the birds' keels were the same color as the desert below me. This fat, less than a teaspoon's worth, had to see them across the sand.

I slept, and then woke even more bewildered over the Congo basin. The vast dry swatch of threadbare desert had given way to a thick sweating blanket of damp dark hairy green forest. To fall into the desert's sand and grit, as thousands upon thousands of swallows and redstarts and sedge warblers must do every autumn and every spring, would be to die by desiccation, the warbler's tongue parched to silence, the swallow's fat cooked to nothing. I thought of the red dust, the color of a redstart's tail, that sometimes blows north from the Sahara and falls with rain back in Britain. Perhaps it brings with it some ghost migrants, the tiny particles of birds that die in the desert.

To fall into forests of the Congo, as many thousands of birds must, too, would be to crash through the lush canopy, then rot and slide in shadow into the dank earth, carried off by mold and maggots. The forest hides more than the desert and is hungrier. Bodies are dusted and rolled in the sand and preserved. Saharan oases are graveyards, heaped with the tiny bones of fallen birds. The forest eats bodies entire, but it also hides the living. Somewhere, in the thousands of feet of wet air between the treetops and me, are Europe's house martins. All cross the Sahara for their winter, but of 290,000 ringed in Britain and Ireland to date only one has been recovered south of the desert. The bird that builds its house on ours, whose selection of our eaves we regard as a benison, whose "pendent bed and procreant cradle" on Macbeth's castle makes King Duncan happy to go inside—all the house martins of Europe are lost to our eyes every winter somewhere over Africa.

We landed. But Africa feels shut, an illegible terra incognita. Once I was sent on a hostage survival course designed to help me cope should I ever be captured while on BBC business. Before our instructors, ex-soldiers, put a gun to my temple and made me beg for my life, they pulled a cotton sack over my head, spun me around a Surrey heath on a hot summer's day, and pushed my clouded face into the soil. Africa does the same.

A kind of continental claustrophobia came over me, familiar from Canada and Minnesota and the Moscow oblast: trees formed the only horizon and gave no sense of ever ending. A hot rain fell. I started to sweat and the sweet smell of animal and vegetable decay didn't leave my nostrils for a week. My body moldered in minutes and lost its compass.

On the hundred-mile journey to Heathrow Airport from my home I had noticed just one species of bird, four matte-black carrion crows pulling away from the mess of a motorway verge like rag-and-bone men. After such a scant, monochrome spread, Africa's birds were refulgent and chaotic. Rainy-season Zambia is wet and warm and green and its birds, mostly invisible to my untutored northern eyes, sounded wet, warm, and green, as well. They showered welding sparks and fizzed with electricity. They sounded like frogs, insects, or children. I knew that *my* birds—those wintering in Africa that I understand—were there, but they were lost in the riot.

I learn a call, but the next time the sound comes I dumbly ask for its name again. I think a pink bird on an emerald treetop a bee eater; Claire corrects me, naming a weaver one quarter the size. After half an hour of listening, I manage to find an African broadbill that has been singing only a few feet from me in a riverine thicket that seems to go on forever. Deep from the shade it burps its buzzing frog-speak and simultaneously turns backward somersaults from its perch, flashing a white marshmallow puff of feathers, like an acrobat magician. An African bird.

At the major's farmhouse, I read in *Swara*, the magazine of the East African Wildlife Society, that a new giant elephant shrew had been described from the Udzungwa Mountains of Tanzania. With orange-and-gray fur and a long snout, the gray-faced giant sengi is 25 percent heavier than any previously known elephant shrew, making it the size of a small rabbit. This family feature, the long and flexible trunk, gave the elephant shrews their popular name until recently. They are now called sengis to avoid confusion with true shrews. Naming a shrew with a long nose an elephant shrew would have been a simpleton's response, the equivalent of calling a giraffe a camelopard. But this nomenclatural and genealogical housekeeping has been wonderfully tripped up by recent molecular research that has shown that these forest-floor rodents are actually more closely related to elephants than shrews. Ancestral elephant shrews were members of a "superorder" or "cohort" of beasts called Afrotheria that evolved in Africa more than 100 million years ago. Elephant shrews are related to hyraxes and tenrecs, elephants and sea cows, and the aardvark.

The elephant-shrew-cum-aardvark, bird-that-might-be-a-frog experience defines all my bird-watching beyond Europe. During another English winter, I traveled to California. As my plane turned out over the Pacific Ocean for a moment and bent on toward Los Angeles International airport, I looked down through another porthole and saw Topanga Canyon. Flying over the ice of Canada, a few hours before, I'd been reading about it and had marked it as a possible place to walk and look for birds. I liked the fact that the Byrds had lived in the canyon once too and made their music down there. I looked into the deep and shadowed clefts and at the sun-blasted ridges. Maybe they still did.

Three days later, I walked the tracks up the spine of the hills and climbed higher for several miles. There were birds but they were difficult. I struggled with the wrentit, a bird whose family was totally new to me and whose compound name both offers and denies help, and with the golden-crowned kinglet, which looks so much like the European goldcrest but isn't. I saw a pair of science-fiction birds, Western bluebirds, their electric blue marking them as birds from the future, or rather from an old color-saturated version of the future: a 1950s bird dreaming of the space race.

After the bluebirds, wrentits, and kinglets, watching hummingbirds fly flummoxed me the most and made me feel foreign at Topanga. I know birds by their flight and when the grammar of flight that I understand is extended or challenged, it is hard to see the flying bird for what it is. Ever since I put aside my *Baby Animals* picture book at age three, I have known that I don't really get on with adult birds that do not fly. Penguins, ostriches, kiwis, and the rest are not birds to me. My prejudiced systematics lumps them with dodos, on the far side of things: they are primitives, throwbacks, dead ends, clumsies, shufflers and waddlers, kickers and swimmers, not flyers.

To see an Anna's hummingbird fly around the small trees in the parking lot at Topanga really told me I was abroad. Here was a bird that flew like no other bird I knew. Hummingbird flight—its zips, and stops, and dives—is totally theirs, could only be described as hummingbird-like, yet I found it hard to think of it as a bird's at all. I couldn't follow it, couldn't anticipate it, and so couldn't really see it. It was a flying bird, and yet its flying was outside what I understood to be a bird's world and so it was out of mine.

Knowing my birds has always been important for my sense of knowing myself. But it takes coming to places like the dusty chaparral of a Californian sierra or the electric-green fields of southern Zambia and not knowing what is flying through them or singing in them to realize that. I am not sure who I am if I don't know how to bring toward me the birds that are around me, their habits and their habitats. John Clare recalled how, in the early nineteenth century as a child on a heath, he'd gone in search of things farther away, in search of the edge of the horizon, and had found himself lost, as I felt at Topanga or Choma: "I eagerly wandered on and rambled among the furze the whole day till I got out of my knowledge when the very wild flowers and birds seemed to forget me and I imagined they were inhabitants of new countries."

To be forgotten before you are even known is how almost all bird-watching is. This doesn't stop me wanting to reach after the birds—it makes it more poignant. Bird-watchers are observers of exits, stewards holding open doors to freer spaces. Though many birds live alongside us, fostering our illusion that they've elected to, we are on our own. The house martins don't need us.

But to be forgotten by the birds of America and Africa *and* to be out of my knowledge there, I was lost and then lost again. In my childhood I would have called the birds of these places "pet shop birds" and left it at that. Nowadays American birds seem to me like quotations or new versions of birds that I know from the Old World, while African birds seem like puddings curdled from some endless bubbling evolutionary milk. The bluebirds' retro blue in Topanga took me back in time. The red bishop's party pom-pom at Choma looked like something on its way to somewhere else.

Rescue comes from some swallows showing themselves to be African and knowable at the same time. Toward dusk in Zambia I walk through more warm wet grass to a fishpond with a dam at one end and a smallish reed bed at the other. Orange light is being ladled like pumpkin soup over the surrounding yellow fields. In front of me a sedge warbler calls, hidden in some low water plants, the same dry harsh note that sedge warblers make while attending their nests in the reed beds at Chew Valley Lake, six and a half thousand miles away. This one, at least, has made it across the Sahara and the Congo. It calls again, then falls silent. Palm swifts scud high into the sky. Below them, slowly lowering themselves to the reeds, are two barn swallows. Two birds barely count as a roost but these two are telling enough.

The swallows' last flights of the day turn imperceptibly from hunting sorties hawking after insects to a final minute of a more sweeping flight of great beauty back and forth over the reeds. Describing in the air the shape of a rope or a rein by which all things might have been tied under the sky, this flight throws over the traces of the day. Now everything must be released to the dark. Swallows do this wherever they roost, but here they are dipping down to sleep through a thick dusk of tropical insects, a warm dew, a virtuoso frog chorus, tiny cigarette-glow fireflies lighting up, great thunderheaded clouds darkening on the horizon and pulses of lightning from far beyond it. And the swallows are sharing their sleep with drongos and hammerkops, babblers and jacanas, all African birds flying to their places through the same air.

I can't follow the swallows down into the reeds but they have mapped themselves into this place. They are at home in this marshy valley of a Zambian farm. What I couldn't imagine before—the disappearance into Africa of the birds I thought of as mine—is happening in front of me. The birds have settled, to sleep in one of the places in the world where they belong. Something of the swallows' last flight reminded me of the seamstress I watched earlier this afternoon, on the verandah of the Manila Clothing Factory in Choma. She pedaled her ancient sewing machine, black with gold inlay, and turned clothes beneath her fingers, all the while peacefully looking through the crowds that thronged the streets and alleys of the outdoor market around her.

As one of the swallows flew over my head on its way to its roost, it sang a full round of swallow song, the lovely purling music that signifies the beginning of spring for me. I wondered if the singer above me had hatched last year and had never sung before now. Was I hearing the first creaking-open of its rusty red throat, like a barn door that it might fly through? Would it fly to a shed on the outskirts of Liverpool or to the Chew Valley and into a farmyard singing the same song? It doesn't matter. I saw the swallows at home here and heard them singing. I could add a swallow winter to my swallow years.

The nests of resident birds helped me too. A nest is the sole fixture—and that only temporary—in a bird's life. It marks the longest joining of a bird to a single place. Just as albumen (egg white, part of the nest's raison d'être) worked as a binding fixative in early photography—the stillest part of a bird being used to arrest movement—so the nest holds the bird to the world. It is an anchorage in the comings and the goings, a sanctuary tied to the earth, a refuge to start from and return to. But the nests at Choma are more complicated still. Claire studies tawny-flanked prinias and red-faced cisticolas (both small streaky warblers) that have been parasitised by cuckoo finches laying eggs in their nests. These, intended as homes and sanctuaries, have become battlefields.

In the old fields at the Major's farm at Musumanene—the big tree next to the river, in Tonga—a prinia or cisticola nest parasitised by a cuckoo finch will earn Lazaro and two other farm workers, Kiverness and Collins, twenty thousand Zambian kwacha (about £3). A prinia nest without a cuckoo finch egg is worth one thousand (15p), a red-faced cisticola, three thousand. A day's wage for a farm laborer is around five thousand kwacha.

The nests are beautiful, domed like grass skulls and tied to the stalk of a woody flaxlike plant by just a few strands of grass twisting out the back. Red-faced cisticolas incorporate living leaves into their nests for protection and further camouflage. Using their beak as a needle, the birds weave fresh leaves onto the dry-grass underlay of the nest. It looks as if tiny green rugs have been thrown over the dome.

The men are incredible nesters: by listening for the parent birds' alarm calls or watching them back to their particular corner of vegetation, each of the three manages to find ten nests or so every day of the three-month breeding season. They look for them at the beginning or end of their hours working in the tobacco fields on the farm or in its guava and mango orchards. Knowing the fields and knowing the birds, they walk to the nests as the prinias or cisticolas leave them—quietly, unfailingly, and directly. Lazaro moves like his namesake, up from the earth as if back from the dead, unencumbered, hurrying toward life, nest after nest.

To find a nest anywhere is a revelation. Arriving at one marks a kind of intense trespassing—the parents fly off, the eggs begin to cool or the chicks are unfed, and the season stops. Walking behind Lazaro, I think of John Clare once more, another consummate field-worker, a farm laborer, gooseherd, egg finder, and nest lover. Nests made poems for Clare. A nest was a cottage, a bower, place of happiness and sex, of nature's nurture. He wrote dozens of poems about them, obsessively recapturing and restating the moment of discovery and of sharing the shelter they offered and the comfort of their tether to a place. Every nest poem is a field note; you arrive in his poem, and he is already there, simply seeing things, without binoculars, close up. He wrote how as a child he would make his own nests with "moss from apple trees" and "bits of straws," thinking "that birds would like them ready made." He wrote how even non-breeding birds like to "nestle" in the autumn. Through his long and catastrophic final breakdown, he continued to write about nests and called what was to be his last poem, written in the winter of 1863–64, "Birds Nests." Like the chaffinch's nest in it, it isn't finished:

Tis Spring warm glows the South
Chaffinches carry the moss in his mouth
To the filbert hedges all day long
And charms the poet with his beautiful song
The wind blows blea over the sedgey fen
But warm the sunshines in the little wood
Where the old Cow at her leisure chews her cud

We drop Lazaro back at the farm workers' compound and children, with scraps of maize meal porridge around their mouths, run toward Claire's truck. Their faces remind me of the rattling cisticola's nest Lazaro showed us. The birds gather spiders' webs to line their domed nests and plait them around the rim of the dark grassy hole like a milky mouth. We say hello to the children. Then we step from a world of hatching and birth into a dark where cuckoos are king, a dark that is at the other end of the story.

The farmhouse where Claire works on her eggs has its curtains drawn. On a metal bed raised to the height of a desk in a darkened living room rests the farmer. I imagine him having been laid there on his iron litter by African bearers after a torturous crossing of the continent, but the major, an Englishman, has been in Zambia for many years. Gaunt and yellow with illness, he livens up at our arrival. His young wife, Royce, smiles and offers gin. The major apologizes for the curtains—"We were being carried away by mosquitoes!"

Claire went to work on the major's farm partly because its ramshackle outer fields are full of prinias and cisticolas and cuckoo finches, and partly because the major has amassed an astonishing collection of bird skins and eggs. The collection has enormous scientific value. In three rooms—his living room, a spare bedroom, and his bird room proper—there are ten thousand clutches of, mostly, African birds. The major has between thirty and forty thousand eggs, every one taken, blown, catalogued, and stowed. It is perhaps the largest private egg collection remaining in the world.

The major's eggs are cleanly blown and the skins are beautifully prepared; the labels on the birds are neat and so is the minuscule writing on the eggs, the registers immaculately kept. The collection is an orderly mass grave, but it also a hecatomb. The epic, African scale of it—the strange, quiet fields of eggs laid out in their original clutches, in ersatz nests of cotton wool, in the starless dark of drawer after drawer—leaves me wondering about all those birds that never became, the unhatched thousands, blown to oblivion out of the single tiny hole made in their eggshells.

For the major, the collection is life-giving. Thinking of his clutches, the sick man rallies, like John Clare half-rescued in his asylum by his memory of nests. The major gets up from his bed and walks, pulling open a drawer to show me his Levaillant's and Jacobin cuckoo's eggs. The cuckoo's secret is plain to see: the deceptions that allow its eggs to be fostered. Some eggs show astounding mimicry, being perfect sky-blue matches for the sky-blue eggs of the arrow-marked and white-rumped babblers that the cuckoos were parasitizing; others are flagrant impostors, the wrong size and the wrong color, but they were laid where the cuckoo could have deposited an ostrich's egg in a hummingbird's nest, because the host bird (bulbuls in this case) couldn't or chose not to distinguish the parasite's unwanted gift. It is a war.

In other drawers I look at the major's set of cuckoo finch, prinia, and cisticola eggs. Elaborately and beautifully patterned, no two eggs are the same. The prinias' eggs have three layers of markings. The background color can be blue, white, reddish brown, or olive green; brownish blotches are smeared over this undercoat; and finally, squiggling lines, often extending from tiny pools of ink, run pell-mell across the egg. They look like miniature globes seen from the other end of space.

All eggs start white. The shell gland in the bird's uterus begins the painting, the colors coming from old blood cells that have been broken down into porphyrin pigments in the bird's liver. As the shell is formed, the background color is deposited. Then blotches and squggles—each egg having a unique pattern—are painted into the outer coating of protein molecules or cuticle of the egg. The cuckoo finch can mimic the prinia's backgrounds and blotches but it cannot—as yet—do the squiggles. Sometimes the prinia notices a cuckoo finch's egg in its nest and jettisons it. The cuckoo finch meanwhile—across millions of years—works on its reading and replicating of the globes beneath it in the prinia's nest. The war goes on. Looking at the eggs I think of the language of coevolution—of how "host" and "guest" share the same etymological root, and that "hospitality" and "hostility" do, too.

Outside, thunder is coming and lightning flashes through the brief Zambian dusk. Inside, the electrical supply is wonkier; the power flickers hopelessly on and off and then gives up. The fridge warms. The television draped in a Union Jack tea towel looks as if it is due a state funeral. The major shouts—"Candles, Julia!"—to his temporary cook and takes comfort by speaking of Churchill and the Blitz. Julia arrives, wheeling a wobbly cake trolley across the carpet.

There are no drapes in the bird room to hide the dead. They are everywhere. Wooden drawers containing two thousand bird skins cover the walls. A huge, seven-foot-tall, freestanding cupboard of eggs takes up most of the middle of the room. The skin preparation table is the only other piece of furniture. Here the birds are gutted and their eyes removed; they are then stuffed—often with weavers' old nests—back to some approximate shape. Perhaps a dead weaver in the major's collection had been plugged with the grass it was hatched in. Perhaps its eggs are in a nearby drawer.

Scientific study skins are not mounted in a lifelike pose, as are birds stuffed in the public galleries of a museum, and don't really look like birds at all. In the skin drawers, the birds lie on their backs, their wings folded tight to their bodies as if they have wrapped themselves up. Each has been skewered in death on a wooden kebab stick. Each has a label tied to its feet. None has eyes, glass or otherwise. Where he has a male and female, the major has arranged his bird skins like the marble effigies of recumbent knights home from the wars, lying alongside their loyal ladies in an English church. Nothing is going anywhere from this room.

Claire and I retreat to the room where she works. Here the major's parallel interest in human genealogy is stored alongside his overflow skin collection and yet more eggs. Next to a pink double bed are three chests of drawers of eggs; the nearest to the pink pillows are labeled "Doves," "Cormorants and Darters," "Auks, Petrels," and "Penguins, Pelicans, Storks, Gannet." On the shelves are journals, food supplies, and tools as well as books and birds: genealogy magazines; skins of a pied crow and a black-necked heron, with cotton wool for eyes, both looking like huge, cumbersome feathery lollipops; some oxtail stock cubes and boxes of long-life "Supa Milk"; a hammer; books on coats of arms, on dormant and extinct peerages; four pairs of binoculars; *Men and Armour for Gloucestershire in 1608* by John Smith; a glass jar of Italian herbs; a plastic bottle of raw linseed oil; an iron; *The Auk* 1970–76; and a "hunting autobiography" by Cecil Aldin, the author of *Dogs of Character*, called *Time I Was Dead*.

Resting on the shelf in front of *Time I Was Dead* is a single large gray-and-black feather like an extravagant quill, the primary of a secretary bird, the large half raptor, half stork of the African plains. The shaft has been pushed through a piece of cardboard cut from a box of mints. Written on the cardboard in a fine calligraphy, as if the feather was its own amanuensis and inscribed its own label, are details of the find: "picked at edge of miombo & guava orchard north of homestead, 10 February 2005, a very local species seldom seen at this property."

I take down a Penguin edition of Defoe's *Journal of the Plague Year* of 1722 as Claire at her desk uses her spectrophotometer to measure the coloration of decades' worth of prinia eggs from the major's collection. He has written on the title page of the *Journal*, noting that Defoe mentions "Sir John Lawrence, Kt, Lord Mayor of London 1664–5." He was, Claire told me, a relative of the major's first wife.

I open the book. It is 1665. The plague is killing London. Grass is growing in the city streets; "death reigned in every corner . . . whole families and indeed whole streets of families were swept away together." Against the odds, some bureaucratic niceties were maintained for a time, and details ("found dead in the streets or fields") were entered "in the weekly bill," but soon it became impossible, almost everyone was dead, almost everything was dead. A cart used to carry bodies to the great death pit in Finsbury Fields, just north of the City of London, is later found in the pit itself, "the driver being dead . . . and the horses running too near it, the cart fell in and drew the horses in also." September was the worst month; between August 22 and September 26, 38,195 died in London. The figure, Defoe is certain, is too low; "great numbers went out of the world who were never known, or any account taken of them."

I think of the major's abandoned fields where grass is rampant, of Lazaro and Claire moving eggs between nests there; I think of grass growing through the deserted ruins of Chernobyl and Pripiat and how I heard of the catastrophe there in April 1986 on my car radio when driving to the halls of dead birds at the British Museum; I think of the bus crash I drove past, days before coming south from Lusaka, that had spilled people and their bags along the roadside; I think of the young mother dying of HIV in the worker's compound the night I arrived in Zambia, and the wailing that followed; I think of the man I saw on a bicycle with an empty coffin strapped crossways over his mudguard riding toward the compound the next day; I think of the beautiful smoked-silver Amur falcons from eastern Siberia and northern China that winter here and that I just saw outside above the fields, and how they remind me of red-footed falcons, their near relatives, that I had seen years ago catching dragonflies around the long grass at the base of a Scythian burial kurgan on the Crimean steppe, and how I wondered whether the great mound in front of me contained buried horses as well as Scythian kings, as Herodotus said it might; I think of the death pits of ashy soil in the forest outside Vilnius where golden orioles sang so loud in May that I had imagined the Jews killed there had been shot with that sweet sound in their ears; I think of the major on his sickbed and his empire of the dead, his eggs, birds, boneyards, and golgothas, and I have to stand up and walk about and seek cold water for my hot face. But in the bathroom the taps didn't work.

The West Country, my neck of the woods, is important to the major. One of the first things he said to me was, "My great grandfather was the sheriff of Monmouth," and parts of his family have had long connections with Gloucestershire. Above the throne of the toilet bowl is a framed picture of some of the coats of arms of the county. I write some down, thinking of soft green English winter colors, of Edward Thomas and Ivor Gurney and their Cotswold poems, of a long-ago girlfriend from a small Cotswold village. Where are they now? And where are all the lords and ladies of all the tiny hamlets?

I can hear the major talking to Claire, his conversation swinging between the difficulties of sourcing black sealing wax in Zambia, necessary for the preparation of his will (I brought some from Bristol for him), and some egging memories, particularly the difficulties of finding red-throated twinspot eggs, since the female "sits like wax." I am enjoying the pileup or meltdown of these wax words when another comes along. The talk moves on to honeyguides, one of the most fascinating families of African birds.

Julia is called to clear the tea things and I fetch further drawers from the major's cabinets. We bring to the dining table more dead honeyguides and blown honeyguide eggs than there are anywhere else in the world.

In my mind, ever since I first read about honeyguides, I have tangled them with the beginnings of humankind, with our African genesis. Poorly known and mostly nondescript, they are the size of small thrushes and related to barbets, toucans, and woodpeckers; all of them (there are fifteen species in sub-Saharan Africa) are thought to behave as cuckoos and parasitize other birds; females puncture the host eggs and the tiny hatchling honeyguides kill any host nestlings that have survived the female's first pass; many honeyguides eat beeswax, and the greater honeyguide has even been recorded "eating candles on the altars of Christian missions"; the greater honeyguide also—and this above all is the reason for my thrill—leads humans to bees' nests so that both may share the honey and wax.

Greater honeyguides work only with people. Having found a bees' nest they seek a person and perch near him and sing to him; if the person follows the bird, it moves ahead, calling insistently until, in the vicinity of the bees' nest, it falls silent and "looks on at the humans." Think of that moment: a bird is looking at you, having shown you some sweet stuff it has found and that you and it can enjoy, if you will only open the bees' nest. It is saddening to discover that around African cities, such as Nairobi, the guiding phenomenon is disappearing. The *Handbook of the Birds of the World* suggests that the loss of this behavior shows that human honeyguiding "is not a genetically deeply embedded, very highly evolved aspect of behavior, but that it is, rather, a relatively recent 'add-on' to the wax-seeking and wax-acquiring habits of this species." I see it more as evidence of what living in any suburb does to our relations with birds. They call to us but we don't notice.

I've seen two honeyguide species—in both Zambia and South Africa—but neither was a guiding bird. But Claire reported an occasion with Lazaro after I left Zambia when their attention was drawn by a greater honeyguide fluttering and croaking ahead of them. Lazaro followed it, whistling back to the bird so that it knew he was coming. At times, when feeling hungry, he is able to summon honeyguides by calling them. Happily, his urban life is still some way off. When the bird stopped, Lazaro started tapping the trees. He can hear honey. Near the bees' nest the bird's calls change, growing higher pitched and with shorter syllables. There was a final talking down from tree to ground, from bird to man: the sweet track had been laid and the honey was found.

The major, with the marmalade cleared away and the egg drawer in front of us, is more interested in honeyguides as cuckoos than as honey suppliers. Listing the clutches he has taken of various host birds with greater honeyguide eggs in them, he says, "I am not a modest collector, as you know, I am a vast collector!" The collection is beautiful and extraordinary. Greater honeyguide eggs are white. The eggs of their hosts vary. The major has six clutches of African hoopoes that honeyguides have laid in; six clutches of green wood hoopoes; four clutches of scimitarbills (treasured but "very smelly and rotten" because the female honeyguide punctured all the eggs); four clutches of striped kingfishers; eight clutches of little bee-eaters; and two very unusual hosts—a southern black tit (its brown spotted egg half the size of the greater honeyguide's) and a capped wheatear (a clutch of two very pale blue eggs with one honey-guide's).

The major's deep interest in the eggs and the lives of the honey-guides and other cuckoo-type birds is striking. He loves the British monarchy and good breeding, is hard on what he calls "mongrel na-tions," and is a tough landlord, yet he is drawn to aberrations and disguises, to imposters and cheats. As we look at his honeyguide eggs and skins he claps his hands to his knees and rocks with mirth telling us of John Walpole-Bond, the Sussex bird man and egg col-lector of the early twentieth century. He is especially animated by the story of "Jock" dressed in his egging kit, looking like "a shabby tramp," stumbling on "a rah-rah party in the English countryside, with a parked Rolls-Royce and blankets spread and girls and a but-ler"; the host of the party spotted Walpole-Bond and beckoned to him, taking out five shillings from his pocket to give the tramp, "then Jock strikes and says in his perfect Wykehamist [someone who attended Winchester school] accent, 'I would rather have a glass of your excellent champagne.'"

It is time to go. Claire and I return the honeyguides to their cabi-net. The power comes back on. I go to put the last of the crumpets in the fridge in the kitchen, where they join a plucked and trussed chicken waiting to be eaten, an unplucked but folded knob-billed duck waiting to be skinned for the collection, and some cockroaches happily licking at the empty egg compartment.

We drive back from the major's farm along twenty miles of dirt road to another farm where Claire stays. It is dark. The last bird skins I pulled out were a drawer of spotted eagle owls. But for their eyeless faces, they were exquisite. Laid head to foot with their gray-barred, dusty silver breasts and richly cryptic gray-brown wings and backs, they looked like silver birch logs stacked for a winter fire. Even dead they silenced the air around them. Driving away from the farm the first bird we saw was a spotted eagle owl stalking the sandy gutter. Its ravishing eyes, two wide plates of yellow jelly, stared and stared into our headlights, the most living thing for miles.

Then we kill something. Nightjars rest on the dirt roads. Like giant moths in a dream they flicker up from the sand and away into the dark as we drive. In quick succession we see a Mozambique nightjar with lunatic orange eyes and a pennant-winged nightjar with weird gloved extensions to the trailing edges of its wings, and then suddenly there is another bird lifting too late from the warm track in front of the car and flying at us rather than away. We stop and I scoop up the mess of feathers of a rufous-cheeked nightjar. As its head rolls between my finger and thumb I can feel its pulse faltering. Feathers fall from it like ashes from a burned-out fire. It is dying in my hand and its warm body cools as we drive on. Its plumage, like every nightjar's, is a fabulous mix of silver, gray, orange, and brown, like the woven patterning of the underside of the world or a carpet containing everything of the earth. Its eyes, like black stars, seem to have come from elsewhere.

February

Cronk

"Why is a raven like a writing desk?"
LEWIS CARROLL

Hoarse with fulfilment, I never made promises.
W. S. MERWIN

D awn is long past and the day is under way. In some versions of
the story it is already drawing to a close. A man, perhaps a
god, stands high on a hill looking out over the world. Two large all-
black birds fly at him from far apart, falling from the sky. They
know him and he is expecting them. One lands on his left shoulder,
the other on his right. They settle their wings and dip their beaks.
He cocks his head first to one and then to the other.

The birds are ravens and they have names: Hugin and Munin.
Each day they fly from the man-god who is called Odin, and later
they return to perch on his shoulders and sing into his ears. We
might not think of them as singers, nor what they sing as a song, but
ravens, the largest of all songbirds, probably have the greatest vari-
ety of calls—sixty-four have been recorded—of any bird at all.
They have much to say. Hugin and Munin tell Odin what they have
seen and heard. The ravens have flown out of the past and back
from the future. Munin is memory and recounts what has been,
weaving the day's happenings into the dark fabric of the coming

night; Hugin is thought and projects what is to come, stitching to-day's dreams into the bolt of silk that is the dawn waiting beyond the night. Odin has the whole world on his shoulders.

There is another story. A man, not a god, throws a raven into the sky out over the water from the boat they have both lived on for many days. The man wants to know what will come from beyond the ocean of floodwaters and over the endless horizon. The raven flies to and fro, then far out of sight, but doesn't come back. It has nothing to say to Noah. He launches a dove. At first she finds nothing to perch on, no rest for the sole of her foot, so she flies back to Noah, who "put forth his hand and took her and pulled her in unto him," as Genesis says, like a bird trapper. On her next flight she came back with the beginnings of a nest, the world in her beak. White doves in this story brought better news than black ravens.

Los Angeles seen looking south from the path to Dante's Peak at Griffith Park is an immense flat glittering plain. In the spring boil of the sun, turning cars at intersections far below flash a scarab dia-mante like tiny beetles laboring in a huge dried-up field. Down there I have watched gangs of starlings and parrots, black and green, making their way through the New World. The starlings from Eu-rope and the parrots from Brazil: emigrants, immigrants, escapees, exotic birds that I grew up calling "plastics" because they seemed so garish or outré they couldn't be real, hustling through the jungle of wires around the low buildings, the starlings clotting into evening flocks, the parrots bouncing old songs from their forest homelands off the tarmac.

Up on Dante's Peak, a still older world persists. A small airplane appeared, dragging a giant poster above the heating city to advertise a new burger with meaty skywriting, the languorous flaps of the banner like the reluctant peristalsis of an overfed gut. Then six ra-vens arrive from all sides of the sky, upstage and down, to eclipse the city with their black plumage, darker than anything. They fly as I imagine them doing so on the first day of their creation, away from Odin, or fresh off the ark. Grave and humorous at once, they seem happy with what they have. Their deep *cronk*-ed raven voice-over seems the only commentary required on everything beneath them. More than a thousand feet above the city, one sails on bowed wings toward the burger plane, then switches off its engine and surfs for a moment, before folding its wings away and rolling left onto its back. It drifts like this for five seconds, gliding upside down: a rectangle of black, a spy plane, a writing desk, a dark mirror of ink. With its back to the city, looking up toward space, the bird throws a single *pruk* to the upper air, before following the roll through and righting itself.

The ravens look the oldest thing for miles and put me in mind of Borges's story "Fauna of the United States," which has a species list like no other: "We shouldn't forget the Goofus Bird that builds its nest upside-down and flies backward, not caring where it's going, only where it's been." Perhaps my upside-down raven was looking back toward Odin Street, near the Hollywood Bowl, which I'd spotted as I'd driven past on the way to Griffith Park, and was remembering the days from the beginning of the world. Or was it remembering La Brea, the tar pits below it, where raven bones have been found alongside the bones of long-extinct giant condors and species with names like "fragile eagle" and "errant eagle" and a hundred thousand other bird fossils as well as dire wolves and American mastodons and Shasta ground sloths, all trapped and then preserved—tar and feathers—by the raven-black pitch that started bubbling to the surface, as it still does, forty thousand years ago.

The angels of another Los Angeles, a skeleton city, assemble themselves from the bone hoard at La Brea. One wooden cabinet in the museum there contains the feet of five hundred golden eagles that were taken by the tar after sticky landings. As I watched the ravens at their air traffic control, I could see a staircase of airplanes descending to LAX stretching behind the birds far to the south. I imagined a stack of five hundred golden eagles on their approach to the dangerous riches of La Brea, backing into the distant yellow sky, the ravens bringing them down from the sierras and guiding them in.

The raven is close to being a worldwide bird. It belongs wherever it occurs. Its song remains the same, and its blackness. Crows are different. Two American crows flew around the observatory at Griffith Park, bending metal as they bounced and sang across the grass, their *prang* calls and black feathers flashing in the sun. To my ears, the American crow yaps; its call is like a twanged wire strung between two cans. It sounds irrefutably American. Where some sort of soil sentimentality in me hears earth stuck in the *caw* of the carrion crow, as it coughs itself over the fields of England, I hear a performance—look at me, listen to me—in the voice of the American crow. Ravens sing the blues wherever they are; the American crow sings Tom Waits singing the blues.

It is close to midnight now as I write in Bristol. A mile and a half from me, a raven, the same species as at Dante's Peak, is sleeping on its cliff nest in the Avon Gorge. Its mate roosts in an ash tree nearby. All the corvids of Bristol, of Britain, of Europe, are sleeping their crow sleep now. They are late to bed and early to rise, but in the thick of the night they sleep.

For an hour before dusk tonight I watched six of the eight British crow species in the space of a few hundred yards of cliff top and gorge air, all of them sharp eyed and wide-awake, active and determined members of the crow nation: magpies steered by their own tail across the short turf; a jay hysterical and clowning through furring treetops; rooks skirting the city on the far side of the gorge moving from pasture to roost wood; carrion crows promiscuous and everywhere like dark smudges on my eyes, one five feet behind me inspecting litter, another two hundred and fifty feet below me on the tidal mud at the river's edge; then in the gorge, making a habitat of its air, jackdaws; and last, the bird I had come to look for, ravens.

The jackdaws and the ravens carry and make the character of the place. The jackdaws have always been in the gorge, and when I was growing up in Bristol, they alone defined it. Only since 1996 have ravens—like peregrins—returned to breed here.

In the seventeenth century, Gondar in northern Ethiopia was the second-largest city in the world. I spent one Christmas there nearly twenty years ago and walked through the town looking at its curious Italian modernist buildings, remnants of its short-lived colonial era, and its fabulous Coptic churches with their painted wooden ceilings covered in black angels. As I crossed the city, ten or sometimes twenty enormous vulturine lammergeiers filled the air at any one time, slicing their vast nine-foot wings and wedged diamond tails close over the city's roofs, their marrow-lusting crocodilian eyes, shaggy beards, and rusting breasts coming as close as the pass of a swallow. Aeschylus was reputedly killed by a tortoise that a lammergeier had dropped on his head; the thought made me draw my head into my shoulder. Elsewhere in the world, lammergeiers are birds of extreme landscapes. The only other one I have seen required a four-hour predawn walk up to the snowline of a Turkish mountain range and even then was miles away in iced blue skies. In Gondar they live happily among people, as, once again, do the ravens and peregrines of Bristol.

At the cliff edge where I watch for ravens, there is a dead tree with a dozen jackdaws perched on its branches. All face out over the gorge; half have gleaned beakfuls of dead grass or dog fur for their new nests, making them look like beer drinkers with frothy mustaches, but their eyes hold their intelligence, their grown-up gray iris with its sober pin-dot black pupil. A ripple passes through the tree of birds and something makes them launch. They open and raise their wings and loosen their perch and allow themselves to be carried out over the void. Their legs dangle beneath them like an airplane's landing gear for a few moments as they fall from the cliff and paddle at the wind, like storm petrels on the sea, catching at an updraft that lifts them in formation, countering gravity and throws them into the sky like black confetti, each a shadow of the other. The blast makes them talk. The *chack* of one bird immediately triggers an answering call, which in turn prompts another to speak on. They make a sound which J. A. Baker describes perfectly: "like dominoes being rattled together on a pub table." They are talking to themselves, and yet it is a conversation.

A second group has melted into the flock and now there are forty jackdaws flung across the air of the gorge. They range high and far out over its gulf; black angels, I think, trying to remember my Milton as well as Gondar, angels who fall only to rise again. Their genius for riding the air seems to show its shape—pockets and waves and channels and funnels—but as I peer at them, the wind watering my eyes, I find myself marveling at their irreducible, specific marking of the sky at that moment. If you try to empty your head and open your eyes and just watch the birds, you see that they are not a metaphor for anything: they are not black angels, they are not Satan and his cohorts, or funereal ashes, or a single organism pulsing and flexing with a shared purpose. They are not metaphors (the Greek for "porter"), they carry nothing but themselves; they are birds (a word so old we do not know where it comes from) flying (an ancient Indo-European word that has spun through all of northern Europe). Birds flying.

To be able to see jackdaws or any crows unburdened is particularly difficult. Because they have come close to us and lived among us, eaten us even, all the family have passed blackly into our minds. Meaning has thickened their feathers. Their shadowing of our lives has blackened their shadows. "The raven himself is hoarse" is how Lady Macbeth puts it. He isn't, of course; she has made him so. But the raven has evolved with us perhaps more than any other bird (or wild bird at least; the chicken has moved so far from its jungle fowl beginnings that it has almost ceased to be a bird). If we are close to any bird, it might be a raven.

Around the world and over thousands of years, human imagination has met raven reality again and again. They *cronk;* I hear a blues. They come around camps before armies clash and it seems to us a sign from the future, as if the birds have already flown through tomorrow and know that some of us will die. The ravens read us for real; they see into our future based on their knowledge of our past.

In Scotland, two ravens are overheard talking:

Ye'll sit on his white hause-bane,
And I'll pike out his bonny blue e'en:
Wi ae lock o' his gowden hair
We'll theek our nest when it grows bare.

Driving through Somerset early one morning under a drab gray sky, I turned a corner near Cheddar and disturbed a raven sipping, with the delicacy of a wine taster, from the eye of a rabbit killed by a car. A hundred yards farther on was a second, this one enjoying the pink goo spreading from another rabbit's split anus.

A newspaper in January 1767 describes a blacksmith from Bridgwater in Somerset—fifteen miles from Cheddar—who went shooting on Christmas day: "On Pallet Hill he espied a large flight of old ravens, fired and killed two, which so exasperated the rest, that they immediately descended upon him, and plied their bills and claws so dextrously about his head and face that notwithstanding all possible care was taken of him, he died last Monday. This may appear strange, but our correspondent assures us it is absolute fact."

Gibraltar is one of the most horrible places I have ever been, but it is also offers the most telling juxtaposition of the sedentary and the mobile, the trapped and the free, of animals close to people and ruined by the proximity, and of animals close to people and as intact as ever. One spring day I walked up to the top of the rock to watch drifting raptors spilling north, migrating into Europe from their winter in Africa. In the blue sky above me was a fractured chessboard of gliding black kites. They seemed not to be moving at all, but hanging and waiting for the turning earth to rotate beneath them. Below them a pair of the rock's resident ravens tirelessly worked to and fro, making the outpost their own, flagging on the kites like their Californian relatives.

I looked down from the top of the rock. On its western flank, where the town is, all was deadly, deathly, dying, and dead. What do the people do there? Tourists drink lager; the shops sell booze and cigarettes. Gibraltar's heroic past is lost somewhere in the shadow of the rock. Down below, the street names, the buildings, the ship-yards and cemeteries are all full of ghost lives that have passed: Ezra Pound went there; Odysseus sailed by; Molly Bloom was from there and remembered its "Moorish wall"; Gilbert White knew his "soft-billed" and "short-winged" birds of passage went through there, but its people today and the people who go there seem broken and vague. Who are they? Spanish? English? Something else?

I had booked myself a room at the Hotel Bristol, seeking some connection that was not to be had. On the pavement outside was a dead swift, a mess of dust and feathers: the genius of escape grounded. As I looked at the dead bird, two raucous, chattering par-akeets flew overhead, either introduced or escaped, the bastard bird children of the rock's human occupants.

The macaques were worse. They make you swear and gasp. Dirty brown, tailless, tired-looking, and mute, they lie around, half-way up the rock, debauched. Stern notices in English and Spanish warn you that they bite and instruct you not to feed them; if you do you risk a £500 fine. As I watched, one fingered his penis with his lit-tle pink hand. At first he seemed cautious, as if he were doing it for the first time, but then he sped up rapidly as if he recognized some-thing half-remembered. In ten seconds he ejaculated. He licked his fingers and then lay down as if dead like his neighbors on a wall above the town. These monkeys, the color of dust, with bums like dried cow patties, are citizens of Gibraltar, feral, neither one thing nor the other, an old idea now forgotten, neglected and barely main-tained; the living dead.

All this against the exhilarating drift of raptors high overhead, making their way across the straits from Africa to join us for a Euro-pean summer. Some are sociable: the great flotillas of gliding black kites, eighty or so in one group, making busy skies. Other migrants are on their own, battling the gulls that rise up agitated every time, wannabe keepers of the rock, exercised by the blocky flight of an Egyptian vulture, a short-toed eagle, and two griffon vultures.

This airway of life pushing north, wild and free but also long mapped and familiar, part of the great breathed exchange of spring and autumn, gives all who take part in it a past and a future. Down below everything and everyone seems stuck, with lager to live on and rum-laced "wobbly coffee" as a treat and revved-up cars but not enough road to get into fourth gear, and the monkeys on the twisting path to the peak, with their eyes drifting off elsewhere once they've clocked you for tidbits or had their wank, glazed and dreaming of old genes back across the water in Africa where the kites and eagles and vultures have come from.

Commuting between the flying birds of prey, the mobbing gulls, the macaques, and me were two ravens. They flew out over the sea toward Africa, looked about—I could see them through my binoculars turning their heads—tracked back, and drifted along the ridge of the rock. At one moment they seemed impossibly distant, able to fly to Africa and to shepherd black kites into Europe; and the next one had pitched beside me and seemed to be peering deep into my eyes. Might I be carrion? Their sable throats rose and fell as they breathed. Like no other bird I know, ravens look at you as if they really have seen you.

Behind the raven shone the sparkle of the straits and beyond that the thinning beaten blue of the Bay of Cádiz and the Atlantic. Ten months before the battle of Trafalgar in 1805, a royal navy sloop called the *Raven* ran aground and was wrecked in the bay. Admiral Nelson's body was carried to Gibraltar, after the battle off the Cabo Trafalgar just west of there. From Gibraltar, news of his death spread north toward Britain.

It was warm on the top of the rock, I felt drowsy, and all these ravens ushered me back to my first visit as a child to London, with pigeons coming to my outstretched palm in Trafalgar Square below Nelson's Column and, later the same day, in my memory at least, a raven on my arm, or even my shoulder, at the Tower of London. We all go back a long way.

Ravens are hardy; they breed early in the year. Many pairs in Britain have nests and full clutches before the end of February.

Every year, I go to the gorge in Bristol to look for them, but twice this February I haven't seen any. Today the glare of the low sun is dazzling and I am on the point of giving up when I spot a peregrine, a big female bird, gliding up the gorge. I cannot turn away from a peregrine. For three and a half minutes I watch her as she rides without a single wing beat, able to pull any wind she wants from the air. Eventually she crosses the gorge and flies at a cliff face from where a raven—one I had failed to spot—leaps out like an angry trash-can liner, a flapping dark cape. The two birds fly close and furious, twisting and tumbling, along the cliff and over the trees. The peregrine pulls away, calling *kek kek kek kek kek*, and lands, hiding in a cage of branches and buds.

The raven flies out of view but a minute later comes back, appearing high above me. Having crossed to my side of the gorge, it has climbed and is as high again above the cliff top as the cliff is from the river, perhaps five hundred feet up in the air. It appears to be coasting farther upward, unflustered by the falcon's swipes. Then, at who knows what trigger, it begins a triumphant tumbling sequence down into the gorge. In two minutes I count forty rolls. Each time the bird flaps two or three deep drafts, then draws its wings to its body and rolls, always to the right, and turns upside down. Its neck thickens and I can see its beak open though it is too far away and it is too windy to catch what it is saying. The next moment it rights itself, back the way it came, except for once in the two minutes when it followed the roll all the way through. As it begins to descend below the level of the cliff top into the gorge's turbulent air, the rolling becomes beautifully rhythmic and almost continuous, a black slalom, a black shuttle passing through a loom. Flying is what I do, it is saying, but this flying, this rolling and righting, this is what I am.

Watching the raven in the wind at the gorge edge, I want to run and flap my arms and roll through the air, my falling man forgotten. I remember what Konrad Lorenz wrote about his pet ravens, especially Roah, who only four hours after Lorenz had acquired him as a fledgling from a pet shop in Vienna in 1929 flew freely after his new friend: "It was an anxious moment when my beautiful and expensive bird sailed onward after overtaking me. All I could do was to call loudly, and to start running at right angles to the direction in which the raven was flying, so that he should see me moving. (All birds respond better to a moving object than to a still one.) I saw his great curved beak turning as he looked back at me over his shoulder, and then he wheeled and came gliding back to me." Later, Lorenz adopts a bicycle to speed up his movements. I also remembered Sir Arthur Streeb-Greebling, the comedian Peter Cook's old character, detailing to his supposed interviewer, Dudley Moore. the complete failure of his lifelong project to teach ravens to fly underwater.

It seems that the raven can take it all, and as I walk from the edge of the gorge back toward the city, I find myself making the shape of the rolling bird with my hand out in front of me and trying to synchronize my best efforts at raven-speak with it, to do some raven rock and roll. I look back but the bird has gone.

March

Feathers and Bones

The bird is dead
That we have made so much on
SHAKESPEARE

It was a fine day and K. felt like going for a walk.
But hardly had he taken a couple of steps before he was in the cemetery.
FRANZ KAFKA

One October, in my late twenties, I spent a week on Bardsey Island, the detached end of the Lleyn Peninsula in northwest Wales. My friend Greg was a warden at the bird observatory there. It was unseasonably mild and there weren't many migrants. High pressure brought clear and navigable skies and the birds must have flown over invisible to us, not needing the comforts of the ground. A wryneck did arrive and lived like a piece of snaking bracken for a couple of days along a stone wall near Greg's house. I spent hours watching it, drawn into its strangeness, the brown migratory woodpecker that twists its neck and looks like an ancient contortionist, with feathering as complicated and as marvelous as those other go-between birds—the nightjar and the woodcock.

Bardsey is known for its bodies. In the warm and birdless days after the wryneck had left, I found myself sprawling in the island's western fields where the earth dips and folds, rippling like ribs of sand on a beach. These waves of turf are evidence either of medieval strip farming, the ghost of human labor, or of something even more substantially human. The island is a graveyard.

Bardsey offers outward-bound autumn birds a stopover before the sea. In a similar way thousands of human pilgrims went there at the beginnings of Celtic Christendom, pulled to the island by their sense of it as a sacred place on the edge of things that had wisdoms to offer the mainland of life. The ruins of an abbey founded by St. Cadfan might date back to the sixth century, and in the fields around, where, I dozed under the warm coconut-scented gorse, large numbers of human bones have been found, evidence for the truth of the legend that twenty thousand pilgrims are buried on the island. Making the journey to die, looking out toward the setting sun, sanctified these pilgrims. I imagined their saintly bodies laid out side by side next to where I lay and across the whole of the western half of Bardsey, as in the calm aftermath of a great battle.

Birds, too, have died in great numbers on the island. The image I had of pilgrim bodies began in a photograph I had seen in the bird observatory of lines of bird corpses that had been killed when they collided with the island's lighthouse. For decades through the twentieth century, because of an accident of its design and position, the lighthouse fatally attracted migrants on clouded nights when the moon was new or young. One night in August 1968, nearly six hundred warblers were killed, and on an October night the same year, a similar number of redwings.

That day in my October on Bardsey, as I lay on my back straining to see migrant birds flying south overhead, I wondered whether I was being cradled by a grave. Could this local curve of the earth be a human valley? Rain clouds were building out in the Irish Sea. A peregrine bolted fast overhead, a bird with a wandering name on an island of pilgrims, though this one was a resident. Behind it, tumbling blackly and brilliantly off the high ridge of the eastern side of Bardsey, was a gang of choughs, rarest and most buoyant of all the crows, mopping at the sky. I sat up to watch them and saw a nun walking through the bottom of the field. She lived alone on the island, tending to the however many thousand souls. She wore chough-black clothes but had armed herself against the coming squall with a plastic cape made from a cut-open fertilizer sack of translucent blue, the color of a Tiepolo sky or a dunnock's egg.

I COLLECTED MY FIRST DEAD BIRD WHEN I WAS FOUR. WE MOVED from Liverpool and left the flat in the big house where I was born—the house with the swallows in the garden shed—for a family house, a red-brick, semidetached villa in a small town in Cheshire. My sister was born in the room next to mine when I was three. I heard her first cry. I saw her jaundiced slipperiness poured into one of my father's green woolen socks to warm her first minutes. I went to a nursery just down the road and later to the local junior school a few streets farther on. There were privet hedges and tarmac driveways, curved suburban avenues and parks with iron railings.

In the park I noticed blackbirds and thrushes on the lawns. They came into our back garden too. I liked the way their wide-eyed sorties across the grass were made with alternating caution and boldness. I began to learn their songs. Put to bed on bright spring evenings, I lay awake, listening to the blackbirds beyond the thin curtains. The songs always seemed sad. I felt pinned to my bed by them. The evenings when my parents went out were the worst. I willed the time to go and the birds' music to wind down to a stop. I knew there had to be silence and proper dark, and that only out of the night-quiet could come the blissful sound of my father's key in the front door, meaning everybody was at home, and all was well, and I could drift into the safe harbor of sleep.

I was a mixture of nerves and curiosity. At school I preferred to sit with the teachers in their staff room at break time rather than risk the playground. But at home I was drawn outside to the garden. Again and again I played versions of the same games—I was a zookeeper, and a little later an animal collector on safari. For years I had only to look up over people's heads to summon my imagined and desired future on the African savannah. I walked home from school as I thought I would walk through the Serengeti.

One thrilling birthday in the red-brick house—my best ever, as it still seems—I was given a wooden slide, a yellow tepee, and a wicker picnic basket. My would-be runaway life began. I had grown jealous of my young sister and had protested by refusing to eat anything other than fish fingers. It didn't work, so I packed a small red plastic attaché case with three pairs of underpants and ran out the front gate, turned right down to the main road, and then turned right again. There, I bumped into the father of Guy, the boy who lived next door. He scooped me up and brought me home. Another day with Guy, I dug a boat-shaped hole out of the black soil behind the garage. It looked like a grave, but we hadn't intended that. We were going to paddle our Cheshire canoe to the sea, but before we finished, Guy inadvertently split my forehead with the spade. I bled into my eyes and held my father's handkerchief pressed to my face in the back seat of the car as he drove me to the emergency room to have the earth cleaned from my head and the cut stitched. I still have the scar.

Around this time, one spring day, a dunnock's tiny chalky skull appeared in front of me, seeming to come up from the earth and into my fingers. I had run from the front door of our house out across the drive and onto a lawn where a rose bed was cut from the green baize of neat grass. A man named Mr. Clare, who lodged in our house, looked after the grass and the roses. The bald oval with its twisted hands of prickly stems was his special garden. He allowed me to come onto his lawn and watch him tend it. The grass was smooth and shone, and when Mr. Clare wasn't there, it was good to lie on and roll over. I loved the sensation of the sun splashing into my eyes from high above and then the cool green shoots and wet earth smells flanneling my face.

I knelt down to the grass, stretched out, and began to roll around the edge of the rose bed. The soil at the base of the woody stalks was bone-meal white. On one turn, as I was rolling, I spied in front of my face the minute dome of a skull. It was one inch long and made of paper-thin bone that globed above empty eyes and tapered at the front to a needle of beak. It was so bleached and fragile it seemed more like an ancient ruined shell or a worn fossil or something even less defined, a wrapped or bandaged space. I wasn't sure whether it had sunk into the soil or risen from it. I can still feel the trepidation with which I pulled the skull from the whitish soil, wondering how much more bird there would be beneath it. Earth had gathered in the cavity of its head and, as I eased it out, a fine tilth of dried soil emptied onto my palm, like sand running through an egg timer. Only a year or so before, I might have tasted that soil, just as my baby sister ate the bread crusts and worm casts she found on the lawn, but I was a naturalist now.

I think it was a dunnock that had been in that skull, but I didn't know. It was a bird's head; that was plenty enough. I took it inside but after that it slips from my mind and I don't know what became of it. I grew up calling dunnocks hedge sparrows, and around this time I had noticed my first one solitarily picking through the dead leaves under the shadow of the roses or along the privet hedge. The soil at the surface where the skull came from was a dunnock's place; its beak would have needled it repeatedly, its feet minutely hoeing it over and over. Its dead skull might have been only an inch from where its living head had been.

One March forty years on, my son Lucian, aged seven, walked behind me, eyes down, along the scummy surf of a sandy beach on Coll in the Inner Hebrides. The wind hammered at us blue and sweet with ozone. I watched great northern divers in summer plumage, black, silver, and serious, riding the roll of the sea like well-keeled naval commanders. Lucian harvested the beach. I turned to see him draping a dead gannet around his shoulders, a wing for each of his arms, its dagger head sticking up behind his. He looked like Eagle Medicine Man, the Crow Indian, photographed by Edward S.

Curtis wearing a dead turkey vulture mounted on his head. The gannet slipped and Lucian pulled it back over his shoulders like a cloak, like King Lear at Dover, fantastically adorned and finding himself simultaneously to be made of the earth, on it, and en route to being in it. A few steps farther and Lucian stumbled like the Ancient Mariner, shrouded in his albatross, laboring to carry the bird he had shot, like a dead baby or a newfound twin.

Wind piled on top of wind and just offshore, above the divers, dozens of living gannets banked up, pointed their heads down to peer through the swell, and tipped forward to stab the water. Two more salted, ruined gannets lay on the beach, deader than seemed possible. It was as if they had died once and then again. Dead skin looks deader than bone. Like bodies exhumed from mass graves in Bosnia, the gannets were not the old dead—the bog people cloaked in their own skins of greasy parchment or a bone-dry unwrapped pharaoh—but the recently dead, with impacted flesh and clothes mashed together, blurred feathers and muscle, the beginnings of skeleton, the body smearing off into the sandy earth. The living neatness of preened feathers had given way to mess. Grimly disheveled, unshaved, drunk, they were both repulsive and something you could mourn.

Their eyes die first; at the moment of death the eyeball shrinks slightly and becomes scored with tiny puckers that blank any reflection. No bird collector could keep a bird's eyes. Carrion flies had arrived and were walking over the gannet's dead cornea below me, tasting it as they went. Through my binoculars from a hundred yards away, I watched my son at the edge of the sea walking toward me wrapped in his dead gannet. I waited for him and asked him what he was doing. "I want to keep this," he said.

Five years after collecting the dunnock's skull, I walked through a March beech wood, bare and hanging from a hillside of the Surrey North Downs, carrying a song thrush's nest in my outstretched hands. The scouring winter had shown me the nest the previous day, within the naked hedge that separated our garden from our neighbor's. I had known about it before, but the wind and the pinched severity of early March sunlight put it in front of me again like an invitation to take it up.

The previous July I had seen the breeding song thrushes going to and from the same nest; I had peered at the nest through the green mesh of summer leaves and branches and had reached into it to feel its smooth mud cup. I had rolled the eggs, tiny pale-blue warm skies, into my palm for a moment; I had felt the nestlings, like shrunken, water-filled leather sacks; and I had watched the fledglings, and even tried and failed to save one of them when it arrived bedraggled and seemingly orphaned in front of me. Now the winter had loosened the nest, unstitched it from the branches of the hedge, so that when I leaned in this time and cupped my hand around its cup, it came free like the dunnock's head. I pulled my booty toward me: a nest curved like the sky, curved like an egg, curved like a skull.

The earthen bowl was intact, fired hard by summer sun, autumn wind, and winter frost; the grass outer weave was wan, sallow, and soft; the dead stems had been drenched and dried and iced and thawed in the exposed hedge. In the cup was an acorn cup, a cup within a cup, a nest within a nest. Perhaps a mouse had used the nest as a larder in the late summer, or a jay had stashed an acorn there in its buccaneering autumn days of manic collections and optimistic burials. Perhaps the mouse had eaten the jay's acorn.

As I lifted the nest from the hedge, it came away easily. It seemed finished, dead even, and so I was sure I could keep it, but as I took my prize toward the nature table at school, it was as if it revived and there were still warm eggs in its cup. Through the woods I held it, crunching over beech mast on the worn path to my classroom, and my cargo began to throb in my hands like a dying star. I remember the quiet under the trees. The fuse-wire-thin goldcrest calls, a single curtailed *pink* of a chaffinch, and my breath. Spring was coming, but not yet.

Above me against the harsh sky in the bare treetops, there were tatty old twiggy nests and heaped squirrel dreys like footballs of leaves stretching all the way down the hill, ten dark clumps through the wood. Edward Thomas's poem "Birds' Nests" (written toward the end of 1914, nests prompted one of Thomas's first as well as John Clare's last poems) is about noticing these winter nests, which, he says, "hang like a mark." This wasn't a rookery. The nests were separate and made by several species of bird: a crow, a jay perhaps, wood pigeons, maybe a mistle thrush. Yet in the wood I felt that every nest was tied to the others. Each operated on each. To see them in the twisting wires of the canopy and to discover these forces was like looking at an orrery, a model of the solar system, and my nest-star joined these others in their universe, and I could feel the suck of their atmospheres as we passed beneath them through the wood.

At the bottom of the hillside, the wood became a hedge along a playing field with a football pitch. Breaking out of the trees, I carried my nest now like a grassy trophy across the chalked white lines toward the school gate. I flushed redwings that were feeding along the muddy track below the hedge; they flew into the hawthorns, the rusty red berries and the birds' conker-chestnut flanks interchangeable in the late winter haze. As the birds landed in the hedge from the leaf-strewn track it was as if time were running backward, their reds and browns making autumn leaves on the trees once more, only then to fall again to the earth as I moved farther away, the seasonal pull of my nest's bundle of summer weakening with every step we took toward the nature table.

Dead birds surround me as I write. Nature tables are not encouraged at schools anymore: the dirt and decay of Lucian's gannet would have raised eyebrows, so it came home instead. Since that dunnock's skull I have always lived among the remains of birds. My parents were understanding. My childhood birthday presents subsequent to the tepee and the picnic basket usually included a dead bird. A stuffed water rail looks over my shoulder; I think it died before the First World War. Next to it is the still-stiffening wing of the rufous-cheeked nightjar I brought back from Zambia. I go to sleep watched over by two huge green bronzewing pigeons from Australia, glassy eyed and awake for more than a hundred years under a beautiful Victorian dome. On my desk next to my laptop is a pot of feathers. On my mantelpiece is a line of skulls: a razorbill, a pink-footed goose, a gray heron, a fulmar, some gulls, and Lucian's gannet, whose bill with a jagged lightning-strike fracture gives away its cause of death; it must have misjudged a dive. Boxes of broken eggshells open like views onto tiny glittering beaches and there have been plastic bags with forensic evidence of crumbling owl and raptor pellets; there was once a shed with shelves of nests like an old hat museum; and there was even, for a time, a tray of carefully gathered house martin splatter like a dripped Jackson Pollock, and a drawer of the soft, ash-ended green cigarettes of goose shit.

I learned a lot from these relics. As a child I was most able to feel the time-before-my-time not from family photographs or from reading books about the ancient Greeks, but from staring at the stuffed birds my parents bought for me or by picking up any dead bird I found. Our pet cats brought in mouthfuls of feathers; a weasel dragged a sparrow from the bird table and hung it in a bush; a baby blackbird, fledged for only minutes, drowned in a butt of rainwater; chicks dropped from nests like uncooked scraps of meat; miles of squashed pulp matted the roads.

I peered at the green woodpecker in its glass case, stuffed and mounted in a lifelike pose, hacking forever at a lichen-covered log, its glass eye shiny but incontrovertibly dead. In my mother's kitchen freezer, I had stowed another green woodpecker that I had found on the path through the woods. At school, a friend rolled a dead goldfinch from his palm onto my desk. He had found it in a gutter, its frail legs and feet folded in death as if it was still securely perched on an invisible twig or was being held by the pudgy fingers of the infant Christ in a Raphael painting. Its serenity in death emphasized what it wasn't any longer—alive, flying, and nervous of men and their machines. The car that killed it must have only glanced it, as it didn't seem to have been destroyed at all; but I wondered what sort of death throes allowed it to tidy its wings away and clench its claws. And what had it wanted to do at the moment it collided with the car?

I sliced the tiny wing of a wren from its corpse, leaving specks of gristle still around the severed joint, and taped it onto an old oblong-shaped computer-programming card that my father had given me. WREN, I wrote on it, and wrapped the whole thing in sticky-backed plastic. The wing looked very dead, utterly contrary to the wren's living, hurrying whirr. I tried to imagine its life before it was hit by a car—that life, which I had never known, stretching back from the bird's death with as much complexity and mystery as my own past.

"I want to keep this," said Lucian of the dead gannet, and I gathered my bird fragments in the same spirit, against their own ruin, seeking the preservation of their specific evidence, but also to stretch the lives of the dead in my head back into their pasts. From my scrutiny of the dead I was better able to understand life. The dead birds were just feathers and dust, and collecting their flightless wings was, I knew, as meaningless as mapping their entrails. Yet still, by being solid and identifiable, they calmed me down, and in death they stretched my mind, up and off, as in life they had flown.

There is a contradiction here. Birds fly away. Their dead disappear. They carry no luggage, whereas I have made birds part of my clutter. A woodcock might carry its young between its legs or on its back, and gulls and gannets must pick up seaweed and other vegetation for a few days of the year to build their nests—gulls in March flying over Bristol with beakfuls of moss look like first-timers in a Chinese restaurant—but otherwise their lives are free of junk. To sew its nest in place a reed warbler in a swamp snips near-invisible threads of spiders' webs from the air. Above the Tamuja River in Extremadura in central Spain, a swallow catches for its nest the floating stray breast feather of a short-toed eagle soaring high above. Nests are made, used, and then left to disintegrate.

Young children, like birds, don't have bags. They are not weighed down with stuff. But this changes. Lucian picking up his dead gannet was a way for him to join the world of the dying. To work out mortality, we gather the dead to us. Like the major in Choma, I have become a "vast collector" and my assembled *ornithologia*, my bird clutter, is a kind of dragging anchor of lists, notes, bulletins, books, feathers, skulls, and nests. In my parents' attic along with my soil samples I found a box of old sandy brown foolscap pages of Bristol bird club reports. When I took down the box I wasn't sure whether it was still the monthly bulletins or whether it had made a wasps' nest, so friable and curved into itself had the pile of pages become. When I touched the outer membrane of a yellowed sheet it gave way and my finger went through months of bird records from Chew Valley Lake in the 1970s.

I can remember days at the lake then but my notes, like the bird club bulletins, have mostly either disappeared or turned to dust. I dream of being able to bird-watch without wanting to take notes. John Buxton described noticing redstarts for the first time in his POW camp in 1940 but didn't write anything then, since he couldn't because, as he says, "I had no paper." Allan Octavian Hume, the remarkable colonial administrator in India, gathered a huge collection of 63,000 bird skins and 19,000 eggs and "had planned to write a vast book on the ornithology of India and had made voluminous notes, but in 1884, while absent from home, a servant sold the manuscript in the market for waste paper." Hume never wrote his book and gave away his collection. A tragedy, but a liberation too, perhaps.

The Young Ornithologists' Club was the name of the youth arm of the Royal Society for the Protection of Birds when I was growing up. It produced a black-and-white magazine called *Bird Life* six times a year. I read every one many times over. Today, just by looking at a cover, I can still remember the contents of entire issues. My ability to do this stops at the point in the 1970s when *Bird Life* went to color. By that time I was already trawling for deeper pleasures, stranger fish: the recherché field identification papers and behavior notes in *British Birds* ("Insectivory and Kleptoparasitism by Pere-grine Falcons," "Studies of Less Familiar Birds, 144-Thick-Billed Warbler," "Turnstones Feeding on Human Corpse") and the bleak surveys of ordinary species, with their odd austere poetry, as de-tailed in *Bird Study* (my favorite of all time: "Where Have All the Whitethroats Gone?"—a tacit and wistful announcement that bird men have feelings, too).

For a few years, Lucian was a member of the current incarnation of the YOC, which is now called Wildlife Explorers. Its all-color magazine is not the old *Bird Life;* it has no place for Lucian's dead gannet, for a start. The articles are enthusiastically upbeat: one of the RSPB's slogans is "Aren't birds brilliant!" There is nothing on birds to be collected or dissected or rooted around in, catalogued or scrutinized on a bedside table last thing at night, or kept in the kitchen fridge.

Bird Life in the 1960s and early 1970s was very different. Dead birds were all over them. "A Collection of Birds' Feet" was among my most read articles. ("A coat of clear varnish will preserve the foot and give it an attractive appearance . . . any 'swaps' that you can get can be exchanged with fellow ornithologists"). An editorial anx-iety about egg collecting may have been behind this: an adult per-ception that having something of birds (in a society that didn't have very much, that still looked and felt rationed) was what would most likely draw children to them, and that egg collecting was, though newly outlawed, a *natural* expression of wonder and curiosity. I col-lected feet as instructed.

Ducks' feet got me to meet Peter Scott, conservationist and son of the Antarctic explorer Robert Falcon Scott. One teenage year I won a duck identification challenge run by his Wildfowl Trust. A diligent student, I was nonetheless already cynical about competi-tions and was struggling to love ducks as much as other birds. That day I had to identify captive ducks in pens, photographs of ducks in flight, and duck quacks on a tape, then answer questions on duck feeding ecology and goose migration. Finally, in a kind of far-out version of pinning the tail on the donkey, I had to match a tray of duck bills to duck feet. I won.

I went up to collect my trophy—a silver goose taking flight—from Peter Scott himself, king of the ducks and sculptor of the goose. It was a strange few moments. I had never revered Peter Scott as I had Gerald Durrell (with whom I'd even had, aged seven, a short but inspiring correspondence on my plans to emulate him) or the young David Attenborough (whose *Zoo Quest* television programs sent me charging out into the back garden setting traps and building cages for hypothetical lemurs and tenrecs). Ducks just weren't very interesting to me and I had also thought of Peter Scott as a bad man for his past as a wildfowl hunter. His paintings seemed catastrophically lurid—it was always dawn or dusk, with wild purple skies doing far too much—and his ducks in flight, for all his lifelong looking (and shooting), seemed no more viably airborne than my granny's ceramic trio. I knew and remembered his father's famous last letter to his wife from Antarctica ("Make the boy interested in natural history if you can, it is better than games"), which kept Sir Peter, no matter how venerable, a child in my mind, and always and only the son of the great hero-failure father. As I walked up to him I thought he should quack my score and his approval of it. As it turned out, he shook my hand and grunted something vaguely patrician, and I thought of the touch of his father on his baby hands all those years before and realized that in my mind he was already dead.

One cold early Easter holiday weekend, when my children were small, we went to Suffolk. We stayed in a bed and breakfast of nightmare chintz with a landlady to match. Lucian was a baby and Dominic, his older brother, a toddler. They were tired but wouldn't sleep. The next day was wet and windy and my mood didn't improve. I was being horrible to everyone and needed to get outside. I grabbed the children, strapped them in their car seats, and, willing them to sleep, drove as if in a hearse through the dripping lanes, the bare trees in the gale shying over us.

The weekend had already been stuffed with bodies. On the drive from Bristol the traffic was backed up, there was a queue, and eventually we passed the cause: a crate or more of chicken portions had slipped from the back of a truck, smearing a hundred-yard slimy mess of chicken, a pink-gray pulp with the odd drumstick still not flattened by passing vehicles. It looked horrible dribbling across the road, and it smelled even worse, partly cooked by the heat of pummeling tires and crudely barbecued by exhaust fumes.

As I drove the children in Suffolk in our holiday hearse, I saw more carnage. Last night's road casualties were beginning their passage into the pulped world. A carrion crow, like its raven brother in Somerset, was going in at the ripped backside of a rabbit, the intact fluffy scut softly stroking the bird's black head as it dipped at the riches it found there. A wreath of cherry-blossom feathers lay on wet black tarmac, the impact scatter of a wood pigeon that must have been hit in flight and bounced onto the road before being tossed clear. A hedgehog was rolled on its side like a knocked-over table, its four legs tensed and straight. It went on and on: a badger, like a giant's shaving brush; a muntjac deer, like an abandoned coat; a moorhen with splayed and hopeless green legs; puff after puff of sparrows.

At last the children slept. I drove and counted corpses. Above the road rooks took chameleon walks along the outer branches of hedgerow trees, carefully fetching twigs to repair and rebuild their nests, intent on their spring. I thought if the children carried on sleeping I might pull off the road and scan for marsh harriers over some seething reed beds and something of the hunting birds' buoyancy might ventilate me.

I never made it. A hen pheasant stepped from the berm and into the road, turned her head for a moment toward me, and disappeared under my front wheels. A ghastly *thunk* reverberated through the car. In the mirror I could see, behind my boys' sleeping heads, the crashed mess of brown on the wet road, one wing hopelessly raised in a surrender flag. I stopped and got out, keeping the engine running, wanting the children to sleep on. Feathers were still coming from under the car, launched back into the air by my exhaust. Dark blood seeped from the pheasant's beak and eyes; her eyelids were half-shut. I could hear Dominic stir in the back seat, but I knelt to the road and assembled the ruined bird in my hands, and carried her to the grass and put her down. Her brown body immediately looked like the dead flowers wrapped in faded cellophane that mark fatalities along dangerous roads. The pheasant's blood and some blown feathers, the impasto of our contact, remained in the middle of the wet lane.

Only once have I seen a bird simply die. In another March on my way to Chew Valley Lake, I watched a black-headed gull fall dead from the sky onto a road in south Bristol. It was part of a flock heading from their overnight roost on the reservoir back toward the city I was coming from. Its falling flailing wings caught my eye as I drove beneath it. The flock opened slightly around it, just enough to let it fall through them as they flew on. None looked anywhere other than where they were going. Dead already, the gull spun down to the edge of the road, where it immediately looked like any other dead bird. I didn't stop.

Thomas Hardy died in 1928 aged eighty-seven. One of the last memories he spoke of to his wife, Florence, was one of his first. He remembered at the age of four or a little older "being in the garden at Bockhampton with his father on a bitterly cold winter day. They noticed a fieldfare, half-frozen, and the father took up a stone idly and threw it at the bird, possibly not meaning to hit it. The fieldfare fell dead, and the child Thomas picked it up and it was as light as a feather, all skin and bone, practically starved. He said he had never forgotten how the body of the fieldfare felt in his hand: the memory had always haunted him."

There are many ways to death and there are no birds there. Virgil chose the Lago d'Averno, a lake in a volcanic caldera near Naples, for his gateway to the underworld. He called it Avernus and it takes its name from the Greek word *"aornos,"* meaning birdless. Birds flying over the lake were asphyxiated by noxious sulphuric fumes rising from the magma reservoir beneath the lakebed. They fell out of the sky.

April

A Singing World

Aria Spontanea . . . An Air that whizzed . . . right across the diameter of
my brain . . . exactly like a Hummel Bee, alias Dumbledore, the gentleman
with Rappee Spenser, with bands of Red, and Orange Plush Breeches,
close by my ear, at once sharp and burry, right over the summit of Quantock at earliest Dawn just between the Nightingale that I stopped to hear in
the Copse at the Foot of Quantock, and the first Sky-Lark that was a Song-
Fountain, dashing up and sparkling to the Ear's eye, in full column, or or-
namented Shaft of sound in the order of Gothic Extravaganza, out of
Sight, over the Cornfields on the Descent of the Mountain on the other
side—out of sight, tho' twice I beheld its *mute* shoot downward in the sunshine like a falling star of silver . . .

SAMUEL TAYLOR COLERIDGE

I magine being able to look down a giant microscope to com-
pare two laboratory slides of cross sections of Britain, one
made in the winter and one in the spring. Think how the edge of
the section grows so extraordinarily when the spring comes. Start
in February when the ravens are breeding in the Avon Gorge, but
all else is locked shut. Take a few square feet of the woods
around their cliff nest and cut vertically down into the earth.
Make a thin slice of what you have cut, your blade shearing down
through the twigs and branches and the trunk of a tree and on
into the earth, through roots and soil to the rock of the crust. Lay

this sliver of world flat on a slide. Draw a fine line around it with a black pen.

Now prepare the same slide anywhere across the Northern Hemisphere in late April. The earth is stretching. The world has a new growing edge. Every twig now holds a hundred leaves, every branch a thousand more, and the black ink line must extend around each leaf as it opens. The surface of the world has massively increased. The suface of the world has ballooned. The black pen must be swapped for a green brush. Add birdsong to this cross section of the earth. Think what the shift from thin winter contact calls to the open-throated full spring song of residents and new returning summer visitors does to the volume of the world. Add the blackcap's fruit juice song in the lime tree above the raven's nest to the biomass, and so extend the line drawn around the world. Calculate the din.

The thought of these slices of the singing earth came to me in a mortuary in a Moscow hospital that I had visited to produce a radio feature about bodies in the Soviet world. On a stone slab I saw a corpse, a man whose purplish fingers and toes and penis looked like wild spring orchids gathered at the edges of his body. Upstairs was a pathology laboratory that reminded me of the kitchens of my elementary school—hot and steamy and full of dinner ladies. Large women in head scarves and white coats were talking over the grind of machinery, while aluminum pans kept stewy things simmering on hot plates. The women prepared slides. With scalpels they cut tiny purplish slices of meat from bigger lumps and carefully poured hot cloudy fat over them. On their tables were cooling bricks of this fat, a sort of wax that became opaque when cold. Other women took the cool bricks and sliced finely at them, revealing specks of meat, tiny bloody clouds set in skies of wax. It looked like the worst possible school meal. They sliced, and each slice became a slide, ready to be scrutinized.

Outside the wooden hospital building, a collared flycatcher sang from a stand of silver birches. It was early May; old and dirty pock-marked snow, dotted with a few new green leaves, lay around the trees. Just back from Africa, the flycatcher sang a thin but strong song, and his silvered notes were thrown like meltwater. This was its wood and it would breed here. Glancing toward it from the pathology kitchen cleared my head. If some accounting were going to be made, even of diseased or dead tissue, shouldn't it take notice of that little black-and-white bird adding itself to the fresh green leaves of the birch trees and sweetly working at the air above with its tiny song? As I left I tried to include it in the slides in my mind. The women were thinking of other things. The bubbling pans on the hot plates turned out to be their lunch.

Spring forces us to notice this effect—this extension of the world. The season arrives as a disturbance of air. New weather seems to have come from far across the sea, new music from deep within the earth. The first flying swallow seen half a mile away cutting through the March cold over Chew Valley Lake gives me hope; the first singing swallow heard adding song to the April sky confirms all that I wish to know.

Birdsong is inherently warm. It thickens the scene and deepens space, adding to the density of places in the lightest of ways. There is more of it in spring and it travels farther than most winter calls. Song allows you to hear more distant things. The blackbird singing on the other side of the park brings the other side of the park into my head for the first time since last May. A greenfinch is singing on the next street over from mine, a street I never go down but that I can now hear.

On a late February day at a suburban crossroads in Bristol, a blackcap is calling its harsh *chack* from a eucalyptus tree. The sound bristles the leaves. The hardy thickset warbler is here at the limit of its possible winter range. Much more north or any more ice and it couldn't survive. Its call seems to assert its hunched endurance. It cannot afford to say very much. Calls are cool, defensive, seeking and searching. They speak of being apart.

The next day is milder. In the same place is the same male blackcap. Today I don't see him, but I hear his curious, inward, whispered subsong coming from deep in the eucalyptus. Subsong is one of the quiet secrets of the world. Almost everyone hears it and almost no one notices it. It is the soft, under-the-breath singing that songbirds do at either end of the summer. Blackbirds and other thrushes do it, robins and warblers. Commonly done from hiding, it is often only heard as you walk beneath or past the tree or bush in which it is being sung. This muted, solitary singing sounds sad to me, as if it isn't meant to be heard, as if the bird is talking to itself or humming slightly self-consciously to cheer itself up, reminding itself of summer.

The day after, the sun shines. The calling and the subsong seem like ancient history; winter has turned to spring in one bird over three mornings. The blackcap is there, bouncing energetically through the crackling leaves at the edge of the eucalyptus, singing as if it has received word of the good things in life. Its song sounds like hot chocolate being poured from an old silver jug. It is last summer's music again, last summer—its fruits and its flies—made music, the same old song but warm and new too. It opens and fills a space beyond the bird and the tree. The caution of winter is over, with its noise that can only reach as far as a breath. The bird's ambitions and energies are extended out into the air, saying, "Here I am, this is me and this is my place, I am staying and you should know."

At the beginning of April, I watch swallows in Shropshire, witnessing the moment they arrive back. Two birds plunge down to the river Onny to drink as if sent from the sky. They seem desperate to taste their river again; surely these are their first sips. I stand on an old stone bridge as they flash up after their skim of the brown stream and pull away from the river toward the barns of the farm just up the side of the valley. Perhaps the bridge is a mark for them to turn left. Do they remember breeding or hatching here last year and recognize the barn? Do they feel the coordinates of the valley, know the taste of its air?

Maybe this male swallow sang on its wintering grounds like the birds I heard in Zambia, maybe it hasn't sung for eight months since it was last here, maybe it hatched last year and has never sung before. But something now, as it flies around the barns, is working like pliers on its beak and out comes a song to match its deep red cut of a throat—a song as if bubbled through blood, or a mouthful of flies, but beautiful.

The farm buildings are changed from one moment to the next as the swallows return. They will be here for months now, flying in and out of the barn and through the yard, filling its air with wings and singing. Their wet blue-slate backs are like a roof after rain, their throats like a single red tile, their breasts and underparts the color of pale stone, their tails streamers like a weathercock's. It is no wonder we think of them as house gods.

The swallows make the place come alive. They fly around with an effortless sense of ownership and appear to be inspecting the farm, ripping and scissoring and plumbing the air between the buildings. It is as if they unpick the farmhouse, the cowshed, and the hayloft, lifting each from its foundations, and then set them back down again, with each given an extra dimension—an airborne configuring—making the buildings settle and sit in a new way.

To see them back—even without knowing the swallows on this farm before—is to feel jolted out of something; their returns define the emptiness of the time when they are not here; their flights make us shiver to think of the sky without them; the rippling start of their song draws attention to the silence before it.

Once I tried to fly myself in a glider, but the weather was against me. Twice I went to the airfield, on the edge of the Cotswolds, leaving my home in sunshine but arriving in cloud. On my first attempt the windsock drooped in listless air. On the second it disappeared into watery mist. With no wind and no visibility the glider I hope to fly in stayed unpacked in its long chrysalis. I sat about in the old shed on the edge of the strip drinking black tea from big mugs. It's an intensely male space, a cross between a tree house and a scout hut. Robinson Crusoe would be at home. The waiting was like the waiting at the ringing station and at the bird observatory on Fair Isle, outdoor men brought indoors to stir milk powder and peer through the windows hoping for a break in the weather. I tried the flight simulator and crashed my virtual plane, so I decided to leave.

I drove off the ridge and down the few miles to the banks of the river Severn. There, the cloud was higher but the rain continued. This was serious, all-day, professional rain. Its deluge after the green burst of the weeks before made the Severn Valley almost tropically lush. The rain turned all life under it to scum. The ground was slushy with fallen things. At Slimbridge, Peter Scott's celebrated wildfowl collection and nature reserve, low and billowing drifts of rain rolled up the Severn like a wet cloth being drawn from the river itself.

The rain had grounded a peregrine, another failed flyer, in front of me. The bedraggled falcon was like a wet cardboard box slumped on the lovely green fields that rose from the river's edge. A second-year bird, it was as big as an adult but streaked and muddy, not yet slate gray and barred.

I sheltered alone in a hide and stared at the bird down my telescope until I felt I was entering peregrine time and space. The horizon—wherever it had gone—should have been the falcon's, but in the rain it stared at the sodden grass right in front of it, its huge eyes moving. It looked sick or injured but was likely just wet and cast down. Waterlogged, its tail dragged in the wet grass, and rain like beads of mercury caped its back and coursed over its head. Droplets gathered on its bare cere and rolled forward, filling the hole on its beak. Every five minutes or so it slowly turned its head and slightly raised its wings, scattering a mantle of rain drops onto the green grass.

The cloud base was at about thirty feet. The peregrine is happy at three thousand. The bird hunched into itself. Cold myself, I felt sleepy looking at it. At the moment when I felt that I could walk right up to the bird, it swiveled its body and seemed to be looking straight at me. How could it tell? I was hidden and there was half a mile of sodden grass between us, but I had been seen. Its huge dark eyes scoured over the top of its hooked beak. Even drenched and spangled by raindrops, this was death's head.

The peregrine turned back and resumed its vigil. I had to tear myself away; the damp was thickening and it was getting dark. When I left, the bird was still there drilling itself into the green, waiting for the rain to stop so its feathers would settle under its preen and it could restore them to the sky. The next day, in a turn in the weather, it would resume its flying and hit something up there with the force of an appalling hammer, and pigeon feathers or gull feathers would rain to the ground.

That same break in the clouds got me up into the air, too. I went back to the gliding club the next day, and as I arrived, the planes were being wheeled onto the long strip of grass on the hilltop at Nympsfield. My palms began to sweat. Having been instructed in the use of my parachute, I clambered into the passenger seat, half sitting, half lying behind the pilot. A passenger to where, I wondered. I felt as if I were lying in a phone box with a stranger. The clear roof of the cabin was lowered over us and everything outside sounded far away.

The launch was straight out of my childhood. It might have been like that to be born, and it was certainly like that at the far stretch of the back garden swing, or at the tipping moment as my go-kart went over the lip of the old bomb crater in the woods, or at the second I let go of the rope hanging from my tree house and tried to swim through the air. These scoffs at gravity bring its answering tug.

A winch wound in a steel rope that had been tied to the glider and pulled it fast and steeply upward. For one second we were rolling along the grass hurtling toward the winch house; for another we were pulled parallel to the grass ten feet above it; then for two seconds we were shot up from the earth, as near to vertical as seemed possible. In those four seconds we went from zero to seventy miles an hour and were tossed a thousand feet into the sky.

I thought of birds I have released back to the air and how the feeling that follows when I open my fingers from around them gives me an intimation of this launch, too: the kick of their feet against my palm as they spread their wings, and their rush upward away from stuff, from my hand and the shadow beneath it, the dark of gravity, into the air and toward the light.

Everything important and desirable seems ahead of us. We didn't ever leave the weather, as a jet plane does, but under the glider's clear canopy I felt as if all the light of the sun was being delivered directly to my head. At that point we crested and the winch cord tightened for a moment, tugging us back, reminding us where we'd been. My ears popped and my stomach pushed at my ribcage.

The pilot released the rope and we were free and flying, gliding down the heating air of the escarpment. The noise of the launch fell back to earth with the zinging rope, like Coleridge's skylark's silvery shit, and then there was only the wind in my ears, the strange noisy silence of the sky. I felt a levelling lurch that was our weight settling, my own weight too, spreading out in the air. My eyes dropped back in their sockets. It seemed like a loosening of life, as if there was a new place to be. But simultaneously came a new sense— my flying sense—my need to stay up, to feel the air and find a way through it. My arms were pinned to my sides but I had a strong desire to spread them wide. I registered every bump that we passed through and found my body bending and adjusting even in the seat, wanting to take account of what the air was doing. It was an embrace, like having a lover's body surrounding mine; where it clasped I held on, where I pushed it replied, where it opened I went.

Then came a sharp tug upward as if we had been hooked by the sky. The plane nuzzled into the air of a thermal. I looked up. Two buzzards had shown us the way. Even though I have watched them from the earth for hours, never before had I felt such companionship as we fell in below them and flew together, soaring around our shared chimney of air. We latched on to a current and began to judder, turn, and rise. Around we went, following the buzzards, climbing through the creamed warm air of the thermal, sucking at its milky way. I think of the African honeyguides calling us in to the sweetness of life.

The light above me and the sensation of the body of the air all around were so engrossing that I hardly looked at what I thought I would notice most, the wind-assisted bird's-eye view of the earth from the sky; but as we came out of the thermal—the buzzards had gone on already—I began to look down. Everything below appeared beautiful but irrelevant. The place we had come from and would return to had no determining power over the life of the sky.

Landing felt like a controlled burying of ourselves back within the local horizon. It was all sensible shoes and mirror-signal-maneuver. The shining map of the earth as seen from the air turned into a dull diorama and rose around our plane like a theatrical backdrop. Never had trees seemed so boring. I missed the birds' horizons as soon as they were out of view, the shadows, folds, and upheavals of all the earthworks, the sunken ridges and furrows beneath the grass fields everywhere, the bulk of the hills on one side of my glider dome, the beaten flatness of the river on the other, and the view beyond, shared by yesterday's peregrine, across the river and into Wales. Even after a first flight it felt difficult to come back down.

As I undid my parachute and clambered from my seat, my eye caught some swallows above the airstrip and the buzzards back up above the wood. My fellow travelers. I saw them as I never had before, as kindhearted acrobats who invited a fall guy onto their stage, as generous keepers of the air happy to let us try to build a ladder to the sky.

I walked away from the planes and skirted the airfield. At the far end of the grass strip was a bird where it shouldn't have been. A lone whimbrel set down on its migration north, its beautiful brown curves streaked back and its fine down-curved bill like the blown brown grass it walked in. The birds normally cross Britain in the spring in parties. As they fly they call a tumbling silvery trill, musical and intimate, that seems to keep them together and makes you look up. But here was one on its own and silent. I felt drawn to the bird, grounded like me after our flights, but it flew nervously into the next field. I had flown for twenty minutes back and forth along the scarp; it had come from West Africa.

I helped to take the wings off the glider. Some of the planes are stored in curious custom-built hangars, adapted sheds that look like they have been made to house a crucifix or an angel. The plane was trundled backward off the strip and the wings, still attached, slid into shallow but wide extensions of the shed to its left and right. The glider was wrapped up. The other storage system allows gliders to be towed from launch to launch. It's more modern but it also looks deathly; the wings are taken off and pinned along the back of the fuselage, and the folded plane is slid into a long tube on wheels like a coffin for a thin giant. Once again I thought of dead birds with their folded wings. And then I looked down and saw one.

At my feet under the transporter, lying on a bier of last year's dead leaves, was a female chaffinch, breast down, her wings folded neatly away. It was hard to know how she had died, but as I rolled her over onto her back there was no question that she was dead. Her eyes had gone to nothing; the feathers of her head and wings were intact but stiff; her breast, exposed by my shoe, was a riot of maggots the color and shape of the glider transporter, which came hurrying out with their legless scuttle from a hole, like the entrance to a tunnel, in the bird's breast. She had been eaten inside out. I couldn't see anything in the cavity. For a corpse less than an inch thick it seemed to have an impossible depth, as if it gave onto another place altogether. A black hole.

AN ELDERLY MAN WALKS AHEAD OF ME DOWN A GREEN LANE WITH spring oak leaves threshing above us. He stops, tilts his head, and raises his left arm. His open khaki great coat follows him as he leans slightly, like one of the oak branches in the wind, and he points, holding up his index finger, at once shushing me and showing me and beginning to count in the music he has untangled from the singing all around us. From the delicious warm thicket of birdsong in the trees and bushes stretching away from the lane, Mr. Rimes has found a deep, fruited moan that breaks and gasps for air, then plunges again into the stream it has made of itself. I have never heard anything like it before.

All of the nightingales that I have heard since this one have somehow been conjured by Mr. Rimes, a short, gray-haired, schoolmasterish man who stowed his sandwiches in his binocular case and wore a tie to go bird-watching. All my nightingales find their way back to him walking ahead of me and my father down a lane in Kent, pointing out the birds singing on either side of us. I have seen and heard far more nightingales in Europe than in England but I still identify the birds' night music as English without quite knowing why: the feel it has of old tunes being suggested and replayed, of jam in an otherwise spartan diet, of the luster of the night sky once London is out of the way, of the ghosts of the Second World War that fill the South East with airfields and small fields with pillboxes and concrete overrun by brambles.

I was twelve. It was June 3, 1973. I know this because a letter tells me so. Among my surviving treasures and relics, my wings and skulls, are a few letters, two from Gerald Durrell (lovingly kept as proof that I could have words with a hero), two from a girl who liked birds (proof of the possibility of such things), a couple accepting my records of spotting a Mediterranean gull and two Cory's shearwaters (proof of my status as serious bird man), and then nineteen others, all in the same wiry but clear hand, all on the same pale blue paper, all with 3p stamps, from Mr. G. B. Rimes of Gillingham in Kent. His letters are proof, mostly, that people used to write letters and that old men shared their boyish enthusiasms willingly: his are almost entirely lists of birds he had seen on his various outings through Kent.

Mr. Rimes didn't have a car. The previous spring my father and I had met him in Suffolk. We were staying in the same Westleton bed-and-breakfast, run by the redoubtable Mrs. Mary Macgilp (a most unchintzy establishment much used by bird-watchers). We'd given him a lift and so began two years of occasional outings with him. My dad and I would drive to his home to pick him up. We would go look for birds, have tea in a café that he had nominated, and drop him back at his lodgings. The only information Mr. Rimes offered about his life in his letters—apart from lists of birds—was two references to landladies. He lived in a boardinghouse, but my father and I can remember no more. Our talk, when permitted in the field, must have been entirely bird driven.

My bird notes from this time are sparse and I cannot now recall if we saw the shore larks we hoped to at Sheerness or the autumn migrants at Sandwich. I don't think we did. Mr. Rimes showed us Cetti's and Savi's warblers (then both equally rare in Britain) in Stodmarsh. Knowing their songs as he did (in fact they are very different, but I didn't know that then) seemed to help him see and I marveled at his ability to coax these birds from a reed bed by identifying them. A little old man in a suit with rheumy eyes and stout footwear was calling up fabulous beasts simply by knowing them to be there.

It was the same when we spotted nightingales along the lane. Mr. Rimes heard the first nightingale and brought it out of the turmoil of bushes and birdsong—the full spate of warblers and thrushes and tits and finches, blackbirds and redpolls, marsh tits and robins. He siphoned its song from the stream. I heard it sing and then I heard it stop, but when it started up again I knew it was a nightingale. Then he managed to produce the bird itself, which he'd spotted between its song flourishes, down on the ground, shaded beneath a bramble, a little rufous thrush with a cocked tail. It paused and opened its beak and I could hear the sound and see the bird twisting its throat and turning its head and its whole body quivering and heaving as it sang.

There were maybe ten male nightingales singing along this lane that day. I have never heard so many in England since. I haven't been back, and I don't know what happened to Mr. Rimes. At the end of 1973, we moved to Bristol and I never saw him again. Nor did I ever discover what his initials stood for.

Though Mr. Rimes and the nightingales of the lane got there first, Keats did something to the nightingale in his ode to the bird—his poetic rendering of the realities of the bird and the ideas it suggests—that is very hard to ignore. His ode does what the best bird poems do. Though it has plenty of "faery" in it, it comes out of real looking and listening. We don't need to know Keats's biography or the details of the poem's making to be certain from it that he knew nightingales. The poem is full of evidence and has the smack of the real, and even—in its lines about the song apparently retreating up a hill—the ghost of a field note. It is not, though, a poem of observation put into metered lines; it wants to think about the bird, too, about its reality but mostly about the thoughts that it triggers. The bird in the poem is rescued from being just a cipher by the way the poem manages to *be* a nightingale, as well as being about a nightingale, and about what hearing a nightingale makes you think and feel. Keats is brilliantly able to mimic the bird's song and mimic the sensation of hearing the song at the same time. It is a double translation out of nature.

Because he has the nightingale's jizz right and we recognize the accuracy of his looking, we don't feel he has hijacked the bird too far from its reality. So when he puts the nightingale to work in the poem, we don't demur because the work seems apt to us. But, we might ask, what comes first: the nightingale or the ode? In the lane was I already in thrall to the poem without knowing it? Has Keats's version of the bird lushly clogged our mind—might his poem keep us from the truth of the nightingale, or could part of the ode's success be the way it asks what the truth of a bird could be?

Keats's poem is a lesson in the cultural construction of nature. Birds are not man-made, but all bird-watching is, whether it ends in the *Handbook of the Birds of the World*, a jar of pickled thrushes, or the "Ode to a Nightingale." The ode spoons a kind of sad green syrup over the nightingale, a medicine that makes you sleepy, not better. An unrepentant Maltese bird trapper and thrush eater walks to his lime twig with its stuck birds, glued by another green syrup.

He has a different nightingale than Keats's in his mind, but nonetheless a humanly constructed one: he is thinking of his dinner.

Yet still the nightingale seems a "poetical" bird. Think of the raven or the kingfisher or the pelican; how slow by comparison is the decay of the nightingale's romantic cultural valency. Keats got the nightingale right and no one has substantially been able to remake it for nearly two hundred years. In this the bird helps, which richly complicates things. Though it sings freely in the daytime and from the tops of bushes in much of its European breeding range, including Britain, it also sings at night and hidden in thickets and scrub. Because it can skulk and because it is a fairly drab, unmarked bird—"no oil painting," my grandmother called it when I showed her a picture of one—its song apparently compensates by being that much richer and more expressive than the song that a bright, pretty bird might need to sing. This helps the song seem like art. So does its suggestion of performance and the way we become observers of this artfulness in a theater—the lights go down and, from some dark corner of a stage, a truth so deep that it can only be heard comes forth in tearing stabs of music. This, too—the actual progress of the song and the way it builds and moves and breaks apart and restarts—seems like poetry or music. A grasshopper warbler's song is no less heartfelt but our hearts cannot feel it, whereas nightingale song seems to speak and we think we might be able to understand it.

Our nightingales are, in this sense, Keats's. Keats maneuvered himself brilliantly between the bird and us and brought us together beneath the same night sky. Nobody has the swallow or the blackbird or the robin as Keats has the nightingale. For a time, I thought that I minded this, thought that the nightingale was tarnished and suffering. I had speeches ready to spring it from its prison and affected to prefer other poetic accounts of the bird (John Clare's "The Progress of Rhyme," which brilliantly transcribes its song, for example), and then I unlearned the poem and these newer favorites, too. I separated my bird guides from my poetry books to prevent any wanton miscegenation and relished science that rendered the birds' song cycles as graphs. I banished all conversational allusion to the poetic bird when in front of the real thing, attempting to record my encounters only in the department of my mind stocked with metal filing cabinets: "May 21, 1977, Inglestone Common, Gloucestershire; wind: light southerly force 2–3; 2230 Nightingale: 3 H."

The "H" means "heard." This antipoetic minimalism didn't work. It didn't need to work. The nightingale doesn't require protection from poets. Finally, I realized I was only doing what Keats said I would. All versions of the bird—poems, musical settings, scientific investigations, plumage descriptions, song transcriptions, dreams—all our nightingales are just accounts, the equivalent of a notebook entry. 3 H.

This thought also occurs in a powerful but broken-down and never-completed poem by Ivor Gurney called "The Nightingales," written around 1925. Gurney had been certified insane in 1922 and was confined to mental hospitals from then until his death in 1937. "The Nightingales" was written in his hand in an exercise book of poems (*The Book of Five Makings*) assembled by him, his editors say, "without hope of publication." The poem has now been published, with its revisions and uncertainties carefully transcribed. Like Keats's poem it mimics the passage of nightingale song through a night, wittingly or not, and the editorial marks across its unfinished lines enhance this. It describes three birds singing. Hearing them, Gurney writes, "How could I think such beautiful . . . it was only bird-song." He wants to resist quarrying nightingale music for meaning and suggestion. But to do this he has to leave the birds. They are too ready-made as poems, too tangled with the gossamer of fairyland and Keats. In order to push beyond the received nightingale, beyond Keats and even himself, Gurney takes another bird from his notebook, which stands in for a nightingale before the poets got their hands on it. A "laughing" linnet sings and it is

As if poet or musician had never/before/true tongue
To tell out nature's magic with any truth . . .

Cleaning his poem and his ears in this way takes Gurney back, and he is able to listen to the nightingales once more. The sense at the end of his poem is not entirely clear and it isn't finished, but he returns to the specifics of the encounter. That we know now that Gurney was remembering and figuring the birds while he was incarcerated in a mental hospital and prone to delusions adds to the poem's devastating effect. But even without this knowledge we can see him, with great care, reaching toward the singing birds and leaving them where they are at the same time, making that beautiful approach and retreat that I think is at the heart of all the best and truest bird-watching—best and truest because it describes what birds do to us. Gurney acknowledges his own loaded entanglement with the nightingales but is chastened by the real thing. And yet the real thing seems poetical, and the birds become poems without being made them:

> *Should I then lie, because at midnight one had nightingales,*
> *Singing a mile off in the young oaks—that wake to look to Wales,*
> *[And]/Dream and/watch Severn—like me, will tell no false/adora-*
> *tion in/tales?*

IN ZAMBIA IN JANUARY, IN A THICKET THAT SEEMS TO RUN THROUGH the whole of Africa, and just after I untangle the notes of a marsh-mallow-frog-jumper or African broadbill, I hear a sprosser, or thrush nightingale, the nearest relative of Keats's bird. It is on its wintering grounds and has found a similar jungle to its breeding habitat in north and east Europe and Russia. It is singing—the song is similar to Keats's or Gurney's—a beautiful mud gurgle, and I am surprised. Is it practicing for Europe or holding territory? I cannot see it, but that doesn't surprise me.

In Moscow in May, it is hot at night and I open my hotel window. At three in the morning I am awakened by fantastically loud birdsong: a thrush nightingale has arrived in the thin strip of roadside birches and is launching great lyrical stabs up into the city night. It is so loud that I think it is a recording. There are speakers hanging in the trees—a feature of the old communist world that was determined to make woods and parks places of instruction—and I wonder if the bird has found a microphone. A street-cleaning truck clatters past like an embarrassed tank; the thrush nightingale is louder. I stand naked at the window looking into the dark trees with my binoculars, remembering Nabokov dealing somehow with nightingales in his autobiography by noting that he heard nightingales while in the toilet. I cannot see the bird.

In eastern Hungary in June, I walk along the banks of the river Tisza. It is hot and sticky. The air is full of willow down and savage mosquitoes, carp are burping at the surface of the water, and the aspens and willows and scrub below are buzzing with river warblers and great warm clouds of thrush nightingale music. I am listening hard, making sure they are not regular nightingales. I cannot see them. I think I have separated the songs of the two species in my mind. I go for a drink of fresh white wine in a cool bug-free cave in Tokaj and accidentally stay for five more. When I come back out into the heat, I sit on a wooden bench at the mouth of the wine cellar and instantly fall asleep. When I wake up an hour later I cannot remember my nightingale song insights. I still can't.

On Fair Isle in September, a loose, scattering wind comes up from the southeast. Fabulous birds appear. Ten steps from the Siberian lanceolated warbler in its tiny field of oats, a vagrant thrush nightingale hops on a garden wall looking like a robin without a red breast. Its eyes are dark. They are eyes for the dark of all the bushes of Europe and Africa and all the night skies between. But here there is no cover apart from a cabbage stalk. I can see the bird. It could be an Italian princess stranded on a northern rock. It is silent.

May

Into and Out of the Hole

There was a tree grew in the wood
The findest tree that ever was seen
The tree grew in the wood
And the green leaves growed all around around around
And the green leaves growed all around
The limb grew on the tree
A branch grew on the limb
A nest was on the branch
An egg was in the nest
A bird was in the yolk
A wing was on the bird
A feather was on the wing
And the green leaves growed all around around around
And the green leaves growed all around

SUNG BY MRS. GRACE COLES AT ENMORE, THE QUANTOCKS, SOMERSET, 1906

But enough of greenwood's gossip. Birdsnests is birdsnests.

JAMES JOYCE

A t eight thirty in the evening in early May on the moors of the Somerset Levels, the sun has gone and the day is cooling fast. It rained earlier, and though the sky is almost totally clear now, the moisture of the fallen rain is drawn back upward and the air dampens toward dusk. Planes drive through the upper atmosphere, their wing lights blinking. Just before the blackout, two swifts, still darker than the darkening air around them, track beneath the planes. I look up because, though it seems improbable, I feel the air being silenced. Flying companionably, together but loose, muted and muting, they disappear into the dark after four seconds.

They are the first I have seen this year in Europe. It is like seeing a seafarer still offshore but returning from the other side of the world. What does a homecoming mean for a seafarer? Where have they been, what have they seen; where will they go, will they ever stop to tell me what they know? The news they carry from the other end of the planet and all points in between is simply themselves. They have been in another air; here they are briefly in mine; that is all. The earth passes beneath them.

The first pricks of starlight follow behind the swifts. The seafaring birds put me in mind of Coleridge, who thought about stars not far from where I am, writing for swifts, too: "everywhere the blue sky belongs to them, and is their appointed rest, and their native country and their own natural homes, which they enter unannounced, as lords that are certainly expected and yet there is silent joy at their arrival."

Chew Valley Lake on May 18, a few miles north of the Levels and two months after my first swallow here. There are thousands of swifts over the water; maybe five thousand, maybe fifty thousand; it is impossible to count. Some are inches above it and others perhaps a mile up. The wind dies and plump drops of rain fizz on the lake surface. Cloud smokes downward and pushes all the swifts down with it. The wet sky is yearning for the wet lake. The air is being squeezed from above, it thickens, and the swifts, thunderbirds, thicken it some more and bring the sky down with them. They are silent. If I shut my eyes I wouldn't know a single one was there.

In these conditions they fly above the water surface in a straight line with deep powered cuts of their wings. Before they reach the lakeside they turn and skim back to resume their passage once more

across the lake. To watch their combing of the matted air is soothing. They don't seem to be feeding, they don't seem to be on their
way anywhere; they are at rest in motion. They have flown from
Bristol to southern Africa and back to Bristol without touching anything, carrying nothing, never perching, never landing, never tired.

Twenty minutes later the pressure eases. The rain stops, the
cloud lifts and breaks. Immediately the swifts want to escape upward
and seem to be pushing the cloud higher. They appear to be able to
run uphill through the air. Some are chasing insects; others just need
more height—they fly up fast and then cut from their climb with a
slicing angled glide.

Those that are feeding use one of two methods. Some explicitly
chase individual insects they have spotted, hawking for them like a
hobby, jinking through the air. Others just fly and collect what
comes to them as part of the huge black shoal of open mouths, siev-
ing the sky like a baleen whale or a basking shark: "from the mouth
of a swift which had been shot, a table-spoonful [of insects] was ex-
tracted."

Some swifts seem content to hang. I can see them minutely adjusting their wings to match the strength of the air they are moving
through, giving them just enough forward propulsion. They are like
a diagram of flight, a model of aerodynamics in a wind tunnel. "As
if the bow had flown off with the arrow," as Edward Thomas described them. Watching them fly like this, making the best shape that
works the air, it seems I can see the wind.

Some are more restive. They begin to fly using an extra-fast
wing beat that flickers quicker than my eye can follow. After these
flurries they glide, some holding their wings stiffly raised in a shallow V. They are displaying. I watch as close as I can and think I can
see some turning over in the air, like ravens, and flying on their
backs for a beat or two of their wings. Ever more bravura flights
begin, with catastrophic dives and wild slingshot accelerations, virtuoso falls and high-speed corrections. The birds—they are a
crowd, not a flock—are working themselves up.

I am watching this seething cloud when I notice one swift twist itself into a headlong dive, pulling hard on its wings as if it feels itself to be going too slowly. A hundred feet below where it started out, I see it pull up in the air, its wings high and flicking above its back. It is alongside another bird, which it has sought out, and it flies closer to this bird from behind and then, in an action I will never forget, it raises its wings still higher above its body and slides itself onto the back of the other bird, which lifts its wings too, and for two and a half seconds they mate. As they do, both birds' wings beat in synchrony extra-deep to their full extent, cutting all the way down and then all the way back up. They look joined in the air, as if one bird has four wings, but it is hard to see whether the male actually rests on the female or holds himself perfectly placed fractionally above her. Then it is over; the female scoots away, apparently breaking from beneath the male, and he banks upward after her, and I lose them in the turmoil of other swifts and in my own exhilaration at what I have just seen. All of my bird-watching life seems to be contained in those two and a half seconds of black magic.

On the Mediterranean coast of Spain, a sudden summer thunderstorm booms and swirls around the old barrio and the castle of Alicante. I shelter in the castle until the rain passes, climbing its dark stone stairways to its outer fortified walls. There on the ground, at the foot of a flagpole, is a pallid swift. Unlike the dead bird in Gibraltar, this one is alive, though grounded and helplessly out of its element. The storm has corralled flying insects and bashed them down toward the earth and this bird, I guess, has followed them but has gotten knocked down by a surge of wind or hail. Its tiny feet and short legs are not strong enough to lift it up; it cannot take off from the ground. I think it is a young bird, hatched just a few weeks before. Its matte-brown feathers are dull and dusty under the electric sky. Since it left its nest cranny, all of its life has been in the air. It has barely needed legs. If it goes back to its nest after it has fledged, it either flies straight in or scrabbles for an inch or two like an amputee, using its wings to propel it. When it returns to the air it simply falls into it, tumbling casually back out into the sky where it lives.

I bend and pick the swift up in one hand, letting its panicking wings fold and calm in my palm. The swift's warmth and its feathery energy feel strong in my hand, as if I am holding repelling magnets made of air. Holding a bird is always a surprise like this. Between my fingers its black eyes reflect the riven clouds overhead. I extend my arm, unclench my hand, and throw the swift lightly up into the sky. For a moment it seems to falter and begin to fall, then, as if reclaiming its magnificent wings from me, it flicks them, dark slivers of moon, and is gone back into its life, airborne and flying.

Every year I keep a diary of the season's last swifts. Or I mean to. You never see them go. Every year they do their disappearing trick once more. They empty the sky of themselves. Out of the blue.

A MAY WITHOUT AN OAK WOOD WOULD BE AS BAD AS AN AUTUMN without migrants. In Horner Wood and East Water Valley on the northern edge of Exmoor, the trees are mostly sessile oaks. They cloak the valleys and rising land as if they have polished it over thousands of years to a series of smooth curves. The moor slopes to the sea along these valleys, and they reach toward it like limbs. Behind them, even in May, the heather on the moor is brown, as dark as dried blood. There are patches of sandy winter grass backcombed toward Dunkery Beacon. But the woods and valley floor are bright green.

Half the trees in the wood have holes up and down their trunk that look like scattered dark eyes. Almost every hole is buzzing with the begging sounds of hidden chicks. As I pick my way across the floor of the wood beneath the holes, sometimes the chicks stop, already schooled in fear; sometimes their pleading increases in volume and intensity, as if they think I have something for them.

On the moor it is windy. As soon as I get below the tree line, the wind becomes music. A timeless tonal roar modulates into something particular to the valley. A green ambush. The hanging trees are arrayed like an orchestra down the hillside, and although almost all are oaks, each sounds a different instrument.

Looking for a wood warbler that I can hear through the leaf shimmer, plying the canopy with its sewing machine song, I peer up through the oak leaves that are still young enough to be translucent. Being beneath them all, the countless thousands of them, millions even, in just a few trees, is like being held within the skin of the wood.

I hear six singing wood warblers before I finally see one. They are the commonest bird in parts of the wood, but they are hard to spot. From below they are the same color as the oak leaves. The birds' washed silvery-green underparts match the underside of the new leaves, and their wings shiver like quivering foliage as they sing. Their color scheme might have been invented in Scandinavia. It is hard to think of wood warblers spending their winters in equatorial rain forests in Africa, since everything about them speaks of a cool spring day on a steep wooded slope in its first temperate green flush. I think of them as intrinsic to the anciently mild wooded valleys, or combes, of western Britain, and yet their pallor and their music has been forged as much in the coppery earth of the forests of Congo and West Africa.

The males' song seems the same color as the bird, smoked silver and clouded green. It is slight and thin but metallically intense. A beautiful descending thread of music, it sounds as if an invisible magician were pulling a silver wire from the bird's throat, like the thin strip of metal in a ten-pound note. The whole bird trembles with the power of its song. Released like a wound-up toy, the males pour themselves into an accelerating trill that reverberates to a halt. And then they do the same again. And again. Every time I hear one I miss the fact that they are no longer silvering the woods on the side of the Avon Gorge.

Below a singing wood warbler is the second of my oak trio. A male pied flycatcher sings from thirty feet up a tree, flitting from one side of the oak to the other. It moves all the time, except for the few seconds of its song phrase, when the muscular effort involved in singing pulls and twists at the bird's whole body so much that it has to stop doing anything else.

Between songs it makes brief flycatching sorties but always returns to the same branch as if rhyming with itself, like its pied body. I sit at the foot of the tree the flycatcher is singing in and lean against its mossed trunk. The bird sings one phrase of song every ten seconds for minute after minute without stopping. The hunting flights, every four or five songs, do not interrupt the rhythm of his singing. He gets out and back in time. His sound is like the small flat stones on the streambed below him being rubbed together, *chiddit chiddit chiddit—drid drid*. As the wood warbler takes on the music of the green leaves around it, the flycatcher breaks open the stones below it, flashing a modest but glittering seam to run through the ore of wood noise. I sit and listen to the same thing for twenty-five minutes. It is beautiful.

In the tree the flycatcher is being clobbered by the wind, but it finds a high bare branch in the sun and stays there, skimming its stony tune across the waves. I watch him through my telescope, his rictal bristles quivering around his beak, his tiny dark eye like a drop of ink, a white blaze dabbed in the middle of his black forehead like a pilgrim's badge, a black back and tail, and a white panel on his folded wings. The wind coming at him from behind has gotten into his white breast feathers so that the neatness of the bird has gone. Every ruffled feather edge is picked out like the teeth of a comb against the deep blue of the sky. Singing all the time, he turns into the wind and the feathers are smoothed instantly. That is beautiful, too.

A hundred steps deeper into the wood are more pied flycatchers. The season for this pair is further advanced. Singing has become a secondary activity. The birds' few days of flirting and display are over, when the male in his crisp black-and-white suit would spread himself on a branch, lower his wings, and shake them as he sang, and the female, brown to his black, was coy but beguiling, allowing herself to be chased through the fresh canopy. All this courtship has been overtaken by the reality of parenting. The male looks now as if he is wearing an old white T-shirt that could do with a wash.

The erotics of the birds' year have shifted. Pairing was musical, mobile, and airy. Creation is milky, fecal, and happens in the dark. The pied flycatchers are nesting in a hole about six feet up an oak where a branch has snapped. The tree is in the hanging wood on the steep side of the East Water Valley. Moss, ferns, and bracken cover the woodland floor below it, where every angle is softened as if cushions have been scattered. Dark green moss has gathered, too, in the furrows of the bark that rise around the hole like the sides of a diminutive volcano. The hole is dark and I cannot see inside.

For half an hour in the afternoon, I record the parents' visits. Twice in this time a nestling appears at the hole and looks out, cheeping open its orange gape. Into its tiny head sunlight presses, and so does the green fabric of the wood, and the river's noise. I can't tell how many young there might be in the dark hole behind it. The adults are very busy. Between half past three and four o'clock, the male comes eleven times with food, and the female ten. At three forty-three she comes twice in a minute. At three forty-seven both adults arrive together and almost collide at the hole. Otherwise, little more than a minute passes without a black beetle, a green caterpillar, or a brown grub being brought to the hole and its mouths. Sometimes the birds fly straight at the hole and disappear so fast into it that I'm not sure whether the male or female has arrived. Seconds later they bounce out again and perch on the small branch just below the hole, as if gathering themselves, as we might pause on a landing of a staircase, before heading off again, up—always up—toward the treetops and their insects.

Whenever they pause on their branch I can look into the eyes of the flycatchers and try to read something of their lives. They are always nervous around the nest. At first I stop too close to it and both birds dither in the branches above, calling until I retreat. There is always caution in their arrival and vigilance in their departure. When the male sings, as he does twice in the half hour, it is both times some way from the hole, much higher up the tree. On the branch close to the hole he calls only *tisck tisck,* and then so quietly I can hardly make it out through the racket of the woods.

Just before both adults arrived at three forty-seven, the earth had spun and the wind had jostled the leaves above the nesting tree to allow the sun to fall directly on the hole for a few seconds. Think of it now. It is May. Picture this green oak wood in southwest England on a blue afternoon; glide over it. Come down from the warm air above the curved crowns of the trees where the buzzards go back and forth, twist through the branches thrashing with leaves to this trunk limed with moss, and fly at the dark hole in the dappled understory. In the time it has taken you to read this paragraph, one of the two birds, perhaps both, will have done this. If it is a May afternoon they will be doing it now. If it isn't, think of these three birds, the male and female pied flycatchers and the one chick that I could see looking out into its new world. Where are they now? Roosting in that same hole for the brief dark of a summer's night; drifting south through the wooded valleys of Devon in August; pushing north through the olive groves of coastal Portugal in March; battling rain to catch flies from an acacia in Nigeria in November; shaking the salt that has dried on their feathers from the sea spray of Biscay in April; dropping quietly from the moist night air of Exmoor back to the same trees they hatched from this May, last May, next May.

Perhaps not. Perhaps there is nothing. Carrion crows have come off the moor into the wood sensing egg yolk and chick meat. They move with exaggerated quietness and stealth through branches and leaves that are too crowded for them. Maybe moments after I stopped watching, this nestling flycatcher was gobbled up or hoicked out of its hole to be stuffed down another's gape. Maybe the male flycatcher sickened and died a day later, leaving the female unable to feed her chicks alone, and the nest became a tomb. Maybe she will drop dead this autumn in the oven heat of the Sahara, the tiny barbs of her mild brown feathers cracking down into grains of bird sand.

I struggle to imagine beginnings and endings in the soft plenitude of a wood fully occupied with breeding birds. To say an animal has no consciousness of its coming death may be true, but it doesn't say enough for me. The flycatchers' darkly quizzical, anxious eyes seem already to have seen it all. Auden's lines, addressed by Prospero to Ariel, in "The Sea and the Mirror" come to me as I watch the birds:

At once from your calm eyes,
With their lucid proof of apprehension and disorder,
All we are not stares back at what we are . . .

An hour later, after walking out of the wood toward the moor, I come off the hill and back into the trees higher up the valley, where the oaks are more scattered and open than down toward the river. This straggly edge of the wood is appealing to redstarts, the last of my trinity. The redstart, the wood warbler, and the pied flycatcher— each sings; all are migrants and carry other places with them; each of them in this English wood was in Africa month or so ago; each makes the scene without stealing it.

Sure enough, I hear a *hooeet* and catch a delicious blur of rust—a female redstart. Her red tail shimmers like the disturbed leaf mold she is picking through on the floor of the wood. She flies and I lose her. Then a few trees away, she calls again, softly. I follow her cautious movement through the branches until she heads straight at the trunk of an oak and in a single action, without stopping, she folds her wings and disappears into a small black hole in the armpit of a branch.

Within minutes I hear eight more redstarts, including two singing males. Redstart song is a dry half trill that starts up over and over, and although it never quite finishes or resolves, it continues to shimmer like their tails. Others call, the wood full of *hooeets* and the thick but quieter *tuck tuck* (like a robin's *tick*, only fatter) that follows. The birds are scared and calling because crows are here, too, and at this edge of the trees where they can move more easily, they are bolder. John Clare called redstarts firetails, and the local name is accurate and beautiful. His sonnet "The Firetails Nest" ends with a couplet that Clare could have written standing next to me in Horner Wood: "Of everything that stirs she dreameth wrong / And pipes her 'tweet tut' fears the whole day long . . ."

I find another nest easily; if I can, surely a crow could too. In a topless birch, whose trunk has snapped off at about eight feet from the ground, is a large cavity shaped like a giant's keyhole or a map of Africa. The trunk is rotting and a winter woodpecker has feasted on grubs and wood pulp and hollowed it out. A redstart shows what he's inherited. Fire is the essence of the male in spring. His back and head are ashy gray, his breast like warm embers, his face black soot, and—best of all—his firetail fans flames of beckoning orange. With a beakful of the smashed wings and broken body armour of a dozen insects, the male bobs and shimmers before leaping into the air, somersaulting over himself, and turning into the hole.

I creep round to look at the nest, moving farther away before I stop, frightened that the crows will watch me watching redstarts. A single chick peeps out, its head still downy; perhaps more are behind it. The mother, the more zealous parent, arrives; in my half hour of watching I saw the male visit once only. She perches for a moment with one foot on either side of the continent, then at the broadest point of the Sahel and the Sahara she moves into their hole. If they survive the crows, she, her chicks, and her mate must attempt to go to the open trees of the wide savannah belt of sub-Saharan Africa, south to eastern Congo and Uganda, in just a few weeks' time.

It's getting late. I walk down through the woods to the river under more nests. A female great spotted woodpecker carrying food arrives on an oak, like a spread fan of playing cards. Her claws, two to the front and two behind on each foot, scratch the trunk and make a drum of the nest, beating out some unique maternal rhythms, drawing her young up toward the disk of light that is the entrance to their hole. They feed and slide back down in the dark and immediately start begging again, panning for food with an electrical *weep weep weep weep weep weep*.

Crossing the river, I can hear a dipper singing. Through the shucking of wet stones comes a weird whispered backward song-thrush song. The bird is midstream on a rock, its feet underwater, the river rushing past it. A dipper is all head; the curves of its neck and back and its short tail are dark brown, like the water-rounded boulders it stands on. It bobs and sings, blinking continuously, its white eyelids flicking like a cataract over its dark eyes, just as the water of the stream breaks over the rock it is standing on. I think of Orpheus's severed head singing as it floated downriver.

A little farther in a shaded back basin, away from the plunge and dive of the river's low waterfalls, the excess of spring has matted into some yeasty fulgence of twigs, leaf casings, cast-off incipient fruits, and dashed-down young leaves. The home-brew scum is thick enough on the water surface for a pair of gray wagtails to walk on it, picking insects as they go, their clouded yellow rumps and undertail coverts giving off a dusty shimmer, their dowsing tails pumping constantly, like a gray brush stuck into a pot of powder paint. Their yellow is another color to add to the wood, a color like all the others: the dipper's brown, the redstart's tail, the wood warbler's breastplate pewter, the flycatcher's T-shirt white—colors already in it, but which the birds manage to replenish and brighten.

I take a steep twisting path on the other side of the river up through the trees back toward the moor. Everything seems continuous and overlapping. Honeysuckle climbs through a holly, another holly has a rowan growing from a suppurating branch, a dog rose is woven through a birch, ferns drip from oak tops. One stretch of the path is strewn with flat stones that thrushes have used as anvils; thirty-eight orange snail shells are smashed along it. The next stretch is an ant path and among the thousands wending their way, twelve are busy like pallbearers shifting a comatose beetle toward their larder. Along the next, a newly fledged robin is trying out ant-ing, opening its tail feathers and wings and squirming them into the earth to encourage ants to crawl and squirt their formic acid, which will kill the lice on its feathers. The last stretch is a sun trap, and an eight-inch slowworm basks so that I can stroke the dark thin pen-ciled line that runs the length of its malted brown back. It makes me lonely and I realize that I haven't seen or spoken to anyone all day. The last slowworm I saw was caught by my sons on a Welsh path like this one. Dominic was curious but cautious; Lucian chased it through the bracken, picked it up, holding on to it even as it severed its own tail, which continued to wriggle on the path for minutes after it was abandoned. Another Orpheus.

By seven in the evening the light has left the bottom of the val-ley, even in late May. I walk up the twisting track at the speed with which the deep cut below me fills with dark, keeping pace with the glare that marked the retreating line of the sun's reach, moving in and out of dazzle and shadow. As I climb I walk through birdsong. The baby bird clamor has stopped, but song thrushes and blackbirds sing, first above me, then level with me, and then below me. Exqui-sitely counterpointed in the evening air, these voices sound as if they are being lowered on the finest of ropes into the valley, to soften the dark for those that remain when all others appear to have left.

I get to the top, where the trees finish and the moor begins to rise, turn, and look back down into the darkening green valleys. The sounds that say you are here, the daytime skyline atmosphere, the radio producer's "wild track," have stilled to nothing. Caves of quiet open to the surface of the earth. We don't often notice this muting of the day, because we are habitually more concerned with the departure of light. Our ancient nighttime fire-making instincts have endured long after we have needed to gather around old em-bers and fresh coals. The light often makes a scene-stealing fuss over its exit as well. Today is a good day to listen rather than watch; there isn't much to see, no beaten gold or bruised clouds, no moon—daylight has simply thinned and gone.

The sound disperses with the light. By nine in the evening, I am able to pinpoint each singer. In the two valleys and the oak woods between them, I can hear six singing male blackbirds. They make a map of song that I cannot hear anywhere without thinking how right Edward Thomas's "Adlestrop" is and how I would send his poem into space if I could as evidence for anyone out there of our happy entanglement with our planet. Even if you don't remember or know the lines, I defy you (or any Martian we might want to meet) not to get what Thomas is describing as soon as he says it, to feel the authenticity of his perception, the way he has noticed something that anyone might, but that he alone has brought to the surface, and how there he has shaped its beauty and made something proportionally beautiful.

> *And for that minute a blackbird sang*
> *Close by, and round him, mistier,*
> *Farther and farther, all the birds*
> *Of Oxfordshire and Gloucestershire.*

The last half hour of birdsong is bewitching. To listen in this way, as the birds stop singing, being able to hold on to the coordinates of the sound, allows me to fill the dark hill and darker valley with quiet life. Because I hear them singing, I know they are there, but, like being at a summer party in a large garden, as the night goes on, everyone gradually stops talking and just sits in the dark.

The birds make a beautiful concert, an ensemble piece to begin with, giving way to two soloists and perfect cadenzas. It is very hard not to hear an elegy for the departing day in what they sing, but before the evening becomes even remotely hackneyed (there being no clichés in nature), there is also a magically transfiguring coda that refuses to let the music finish and begins it all over again, but this time in the key of black.

The two star singers are song thrushes and blackbirds, the song thrush a diarist and the blackbird a lullaby singer. The song thrush's song is an account of what has gone on, a record of struggle, hardship, fear, joy, and passion. The blackbird knows this—we remember its detonating alarm rattles—but its singing urge is to smooth, to take the heat out of the day and to lull it to sleep. The song thrush's song is a haw, but the blackbird's is a ripe blackberry. Evening blackbird song seems to have more rubato than morning song, more complexity and liquidity. Older blackbirds sing in the evening more than in the morning (dawn song is more likely to be the underdeveloped, repetitive song of younger males). We are hearing the wisdom of years at the end of the day.

In this context the third song, the nightjar's, is hardly a song at all. As I prepare to leave for my bed it starts churring behind me and doesn't stop. Night has fallen, it says. Fallen, we say, but the nightjar's song on the edge of the moor—like the grasshopper warbler's on the fen and the woodcock's grunts above it last June—seems to have come up out of the earth and brought the night with it. The sun falls and the nightjar's churring rises, like the moon—except that this night there is no moon, only a black vibration from deep inside a bird. Nightjar song is a black stream of stout pouring from a tap, switched on to run at full pelt instantaneously, switched off equally abruptly. It goes like an engine, yet it is one of the most pre-industrial sounds you can imagine. Its motor is the underworld of the earth.

The nightjar's churr belongs with moth-speak and bat-talk—darkness audible. Where the grasshopper warblers' reels are too high, the nightjar's churr is too low, somehow mostly beyond our hearing, something deeper than we can hear. We hear the rippling burr, the hard trill—what Thomas Hardy called the *scurr* and John Clare the *croo*—and we can even imitate it. Think of the hell of a road drill. Vibrate your tongue against the roof of your mouth and top teeth. Don't stop. We can make the nightjar's sound in our heads, but we feel it in our chests, we register it low in our bodies, as if we are only getting the top notes from a deep shaft of furred bass. Somewhere down there is the full nightjar.

Gilbert White *feels* a nightjar singing when drinking tea with his neighbors in a straw "hermitage" in Selborne, his village, surrounded by beech woods, heath, and farmland in Hampshire, in southern England. The bird lands on the roof and begins to churr, giving a "sensible vibration to the whole building."

The nightjar's song, its superbly cryptic feathers and its nocturnal life remove it from us. But it is still there—singing in the dark and extraordinarily colored. Imagine the most difficult jigsaw puzzle in the world. It is called A Sleeping Log Dies. It is made from the backs of a nightjar, a woodcock and a grasshopper warbler. The nightjar section is the hardest of all—the hardest to describe and the hardest to do.

This is how the great handbook of European birds, *The Birds of the Western Palearctic,* attempts to capture a nightjar in words—one bird sits still among all these feathers:

Forehead, crown, and nape finely vermiculated pale cream or greyish-white and grey, feathers on central forehead, crown, and nape with black pointed shaft-streaks up to 3–4mm wide, shaft streaks narrower towards side of crown and faint or absent on lores or above eye; shaft-streaks on central crown bordered by cinnamon-buff and grey vermiculation; top of head appears buff-brown with bold black streaks on centre, almost uniform pale grey-buff or pale grey on line from lores over eye. Patch in front of eye, sides of head, lower cheeks, and chin narrowly barred or speckled cinnamon-buff and greyish-black; a narrow streak below gape to below eye uniform cream yellow or white; black bristles bordering bill; indistinct half-collar from below and behind ear-coverts to sides of hindneck cream-buff or cream-yellow, boldly spotted black, sometimes extending across hindneck as an indistinct black-and-cinnamon-bared band.

That was a nightjar's head. *The Birds of the Western Palearctic* goes on in this vein for another 160 lines at least. The text is indicative of the whole book. This is the Great Wall of twentieth-century bird literature, eight volumes redundant before they were even finished, but making a barrier still in the mind, a neurotic encyclopedic terminus. I prefer another description—one that seems no less evocative of a nightjar. My reading about the birds' diet took me to a list of moths. Lepidopteran naming struggles as nightjar description does, but the accumulated effect of some of the several thousand marvellous attempts at moth-capture gives me at least an intimation, the *feeling,* of a nightjar. So maybe one hides somewhere here among these moths: the beautiful carpet moth, the green carpet moth, the water carpet, the dentated pug, Manchester treble-bar, lunar thorn, the traveller, streamer, welsh wave, dingy shell, common wave, scorched carpet, pigmy footman, magpie, feathered beauty, dotted rustic, toadflax brocade, mullein, heart and dart, uncertain, exile, scarce dagger, cameo, delicate, Hebrew character, stranger, dingy footman, ghost, common swift, goat, lobster, poplar kitten, satellite, brother, northern eggar, heart and club, flame, gothic, feathered ear, white colon, pale shoulder, double kidney, beautiful snout, Berber, Latin, figure of eight, cream wave, festoon.

The moth names are poetic and beautiful and the scientific text is accurate and as clear as it can be, but in the end neither really helps us see the nightjar. Both lists are moving monuments to human ambition—the wonder and challenge of trying to translate what you can see into something you can understand. Both fail. I love the moth names and I love *The Birds of the Western Palearctic* for "vermiculation"—a word I have only encountered in these and other similar pages of defeat. It describes the wavy lines of color and patterning that refuse any rigidities or edge and allow the bird to merge with its surroundings. All birds, in this way, are vermiculated; and we are not and can never be.

In July 1969, when I was eight, my father took me on my first group bird outing: a nightjar evening on a Surrey heath run by the local bird club. It was a bizarre and magical night. Apart from us, the participants were head-scarved ladies and country-suited gents. The leader, an elderly man, fully tweeded, bald, and with an ex-army jut to his chin, led us through the gloom and between gorse bushes out onto the sparser heath of the common. The night was moonless. We had arrived a little late and had hurried to catch up with the group and joined the already moving dark cloud of scarves and shooting sticks. Silently we tagged along, not knowing what was coming.

I had never seen a nightjar, hardly knew what one was. In my *Birds of Britain and Europe* (the only field guide then: Peterson, Mountfort, and Hollam, eleventh edition, 1967; dedicated "to our long suffering wives") the nightjar was lodged at the bottom of plate 46. Above it were the roller and bee-eater and hoopoe, all fabulously Mediterranean with hot aureate colors that were not, even in the late 1960s, permitted in Britain yet, except under license to the kingfisher. They were the birds I longed to see, but beneath them was a half bird half log, looking legless, dozy, drunk even, really legless, with drooping near-closed eyes. The text opposite said: "Nightjar—'Dead leaf' camouflage above, closely barred below." That was as much as I knew.

On the common the whispering platoon shuffled to a halt. At the back, in the dark, we bumped into the stalled tweeds and scarves invisible in front of us, mouthed our apologies, but then fell into a deeper silence as all around us from nowhere and everywhere the heath began to purr. The nightjars were churring their earth song. One had set another off and the common reverberated. Our leader stepped away from the gang of listeners and took a folded handkerchief from his breast pocket. He flapped it open and like a surrendering major waved it through the night, a small white flag over his head. Suddenly, a nightjar was snapping its wings above its body with a feathered whiplash—a male had been drawn to the white of the handkerchief and flashed his rival white wing and tail spots right over the head of the old bald man, a bird conjured from the night.

Finding a nightjar's nest is very difficult. John Clare, the best nester north of Zambia, never found one. They don't really make one, simply pushing a scrape into a bare patch of heath soil. You cannot separate the sitting bird from the mess of heather stalks or bracken or pine saplings that it is concealed among. The eggs and the nestlings look the same as their surroundings. And in the day they all sleep.

Looking for nightjar nests to help a research project in the sandy heaths and pinewoods of Breckland in Norfolk, I went out with thirty other people and we tramped for six hours backward and forward beating the ground with canes like a blind army, through heath land, bracken, and foxgloves, stumpy leftovers of cropped pine woods and prickly new plantations. Though the area is a stronghold for nightjars, we managed just a single nest. A few yards from me another volunteer flushed a female off two chicks. The bird stumbled up from the ground into the air, her flopping wings feigning injury to distract us, and collapsed a hundred yards away back into some scrawny grass. I didn't see her before she flew and have still never seen the European nightjar on the ground. Because the bird is rare and protected we weren't allowed to look at the nest but were led away and shown a photograph of the chicks instead. Their eggshells were removed from the nest to reduce its visibility to predators. Badgers and roe deer have been caught on camera eating nightjar eggs. As a consolation I was allowed to keep a fragment of one of the shells. It is here next to me now. The day the Bastille was stormed, July 14, 1789, Gilbert White was given a clutch of two nightjar eggs; they were "oblong, dusky, & streaked somewhat in the manner of the plumage of the parent-bird, & were equal in size at each end." My shell is dusky and streaked too.

I was happy with my fragment, although I was disappointed not to have seen the bird break from the ground. Nonetheless, something was right about not seeing it. The nightjar doesn't want to be seen in daytime by anything. I remembered holding the dying rufous-cheeked nightjar on the sand road in Zambia and seeing its eyes shut against the world. It seems I have become a bird-watcher content not to watch birds.

But the same night in Norfolk, we caught a Breckland nightjar in a mist net as it hunted for moths. By torchlight I saw its living black globes for eyes, the bristles around its wide mouth, its moth-slaying moth face, the fabulous tasseled and brocaded smoking jacket it wears, its night mufti, the colors of tobacco juice or a dhurrie trodden into the dusty floor of a shuttered summer house. It was a bird unearthed. And I wanted never to stop looking.

A male, it was banded and released. For a second the white patches on its wings and tail hung in the air, like moths in a torch beam, and then drifted as if the thin half light of the moon and night sky couldn't keep up with their brightness. It vanished. There is no keeping these birds. In August 1786 Gilbert White had a young nightjar "for several days in a cage, & fed it with bread, & milk. It was moping, & mute by day; but, being a night bird, began to be alert as soon as it was dusk, often repeating a little piping note. [I] sent it back to the brakes among which it was first found."

We walked back through the dark along the sandy tracks of the Brecks. The air stroked our faces. The stars were bright and clean far above us. A woodcock flew over fast, grunting through its open beak. I remembered, years before, watching red-backed shrikes a mile or so away, perhaps the last pair to breed in England. And nightjars came flying around us, calling and flashing their handkerchiefs, and glowworms lit their "living lamps," as they did for Andrew Marvell's mowers, along the side of the path, burning green holes in the world, ways out, ways in.

Afterword

Singing

The wild duck startles like a sudden thought . . .
JOHN CLARE

When an individual [passenger pigeon] is seen gliding through the woods
and close to the observer it passes like a thought, and on trying to see it again
the eye searches in vain; the bird is gone.
JOHN JAMES AUDUBON

In November 1971, when I was ten, a girl kissed me for the first time. I couldn't kiss her back; a mistle thrush got between us.

A cold ragged day had begun without promise. The year had pulled into itself. Light came up but there was no sky; the blank space above looked as if it had been rubbed with a dirty cloth, a worn gray smear pushing over everything. It was Surrey and the weekend and I was bored. Autumn was over, Christmas a long way off.

My mother called me to the front room and pointed through the window. There was a girl standing on the road, at the bottom of our driveway, below the line of bare beech trees, looking up at the house. I knew her—she was named Karen—and I knew immediately why she had come, but I wasn't ready. My mother said she had already been there for a while and that I should go out to her. I didn't want to but I did.

I opened the front door, starting down the gravel. It was about a hundred yards to Karen. As I walked I could see her lips moving, her mouth opening and closing. I couldn't hear her because a mistle thrush had started singing its song from somewhere high up in the trees.

I heard it as you might "hear" a lighthouse—a voice on its own, powering away through the wind, a clean shout of far-carrying pure song. It lit all quarters of the sky space with short repeated stabbing notes that made me wince as if it were cold.

Karen had come up through the hundred-acre hangar of trees—the same wood I had walked through holding my song thrush's nest—from where she lived, near our school. She was like a mistle thrush herself—lean, leggy, a little severe, with a short haircut and spectacles—and as I stepped closer, though I still couldn't hear her, her lips moved and she seemed to be singing the thrush's song, as a troubadour might recruit an "auzel." Love had brought her here, love of a kind, like the thrush. She and it were working against the season, it was November, the days were still shrinking back, but the bird was lighting the way to spring and Karen wanted to do the same.

I couldn't answer her. I was not even eleven, she was six inches taller than I, and it was all too soon. The thrush had stopped my ears. Karen smiled and tried again. She looked down sweetly at me, her head falling to her chest as if she'd been hung from a hanging tree. She steeled herself and said that she would go if I would let her kiss me.

She leaned in and kissed me on the lips. The mistle thrush was directly above us, high up in the broomhead of the bare trees. I felt it was singing into my skull, annealing my whole body with its bright white music, heating me up and cooling me down. I followed Karen's eyes as she looked up and replied with a peck on her cheek; it was all I knew. She turned and as I watched her cross the road and start down the path through the wood, the mistle thrush was still banging on, shouting after her.

The same afternoon my mother, father, sister, and I visited a traveling funfair that had stopped on a playing field in a village south of where we lived. At one stall my dad and I tried to try to throw Ping-Pong balls into glass bowls. One of his shots went in and we won a goldfish, an inch of orange peel in a clear plastic bag of water. Somehow at another stall I won a beautiful globed balloon, silver and round like the full moon and filled with helium. In the car, on the way back, I was happy; Karen and the thrush were somewhere in my mind, but not at the front of it. I played with the balloon, loving its buoyancy, its urge to go, like a dog on a lead, to tug away from me toward the roof of the car. I held the fish in my other hand, turning the bag around it while it stayed still, facing me, unmoving and hanging in its pint of water. At home, I hurried toward the back door to take my two prizes inside, but as I ran under the washing line in the garden, the string snagged it and I instinctively let go to free myself and in an instant the balloon was released and lifting, the string lurching up, too high already to catch, and worse—I don't remember doing it, but at some point on our journey I had tied the string to the handles of the plastic bag—it took the fish with it. The balloon and fish lifted fast, up over the lawn and the roof of the house, where the wind took possession of them and bumped them across the opening sky, back towards the road, getting higher, not just above our garden any more, but over our neighbors' as well, and higher still, clearing the beech trees where the thrush had sung, the balloon's silver and the swinging goldfish tiny like a toy, like a cutout in a paper theater against the twilight, so small, and going, but going on and, strongly, away from me.

> *The palm at the end of the mind,*
> *Beyond the last thought, rises*
> *In the bronze décor.*
>
> *A gold-feathered bird*
> *Sings in the palm, without human meaning,*
> *Without human feeling, a foreign song.*

Notes and References

This running sky —Philip Larkin, "If hands could free you, heart," in *Collected Poems,* ed. Anthony Thwaite (London: Marvell Press & Faber and Faber, 1989), page 294. Hereafter "Larkin."

INTRODUCTION: FLYING

these were the first words—Louis MacNeice, "All Over Again," in *Collected Poems,* ed. Peter McDonald (London: Faber, 2007), p. 572. Hereafter "MacNeie."

"Here at the fountain's sliding foot"—Andrew Marvell, "The Garden," in *Andrew Marvell,* The Oxford Authors (Oxford: Oxford University Press, 1990), p. 49. Hereafter "Marvell."

JUNE: BLACK BIRDS AND WHITE NIGHTS

"I will sing of the white birds"—Ezra Pound, "Cino," in *Selected Poems 1908–1969* (London: Faber, 1977), p. 16.

The seagull has no English—David Thomson, *The People of the Sea: Celtic Tales the Seal-Folk* (Edinburgh: Canongate, 2001), p. 81. Hereafter Thomson.

a boy caught a moribund black-capped petrel—William Yarrell, *A History of British Birds,* 4th ed., revised by Alfred Newton and Howard Saunders, 1871–74, vol. 4, pp. 8–9. Hereafter "Yarrell."

forty storm petrels . . . were forcibly displaced—Norman Elkins, *Weather and Bird Behaviour*, 3rd ed. (London: T. & A. D. Poyser, 2004), pp. 216–17.

two thousand Leach's storm petrels—E. M. Palmer and D. K. Ballance, *The Birds of Somerset* (London: Longmans, 1968), p. 46.

JULY: NEITHER SEA NOR LAND

The full moon glided on—Samuel Taylor Coleridge, notebook entry, Cumberland, 31 October 1803. Geoffrey Grigson, ed., *The English Year*, (Oxford: Oxford University Press, 1984), p. 145.

Rusham Road, 6–7.15 a.m. one thrush hammering away—Edward Thomas, notebook entry, Balham, London, 8 July 1913. Quoted in Edward Thomas, *Collected Poems*, ed. R. George Thomas (London: Faber, 2004), p. 213. Hereafter Thomas *Collected Poems*.

big humming ewer—Gerard Manley Hopkins, journal entry, Lancashire, 16 June 1873, ed. Gerald Roberts, *Selected Prose* (Oxford: Oxford University Press, 1980), p. 57.

"The poetry of earth"—The opening line from Keats's poem "On the Grasshopper and Cricket," in *John Keats*, The Oxford Authors, ed. Elizabeth Cook (Oxford: Oxford University Press, 1990), p. 53. Hereafter "Keats." The poem was written on December 30, 1816.

"cockeroades"—See J. H. Gurney, *Early Annals of Ornithology* (London: H. F. & G. Witherby, 1921), p. 192. Gurney quotes George Owen writing in 1603 on woodcocks in Pembrokeshire: "Their chief taking is in cockeroades in woods, with nets erected up between two trees, where in cock shoote time (as it is termed) which is the twilight, a little after the breaking of the day, and before the closing of the night, they are taken, sometimes two, three or four at a fall [of the net]. I have myself oftentimes taken six at one fall, and in one roade at an evening taken eighteen . . ." Hereafter "Gurney."

"quorr quorr-quoroPIETZ"—Stanley Cramp and K.E.L. Simmons et al., eds., *The Birds of the Western Palaearctic, Handbook of the*

Birds of Europe, the Middle East, and North Africa (Oxford: Oxford University Press, 1977–94), vol. 3 p. 454. Hereafter *"BWP."*

"the woodcock with her long nose"—John Skelton, "Philip Sparrow," in *Selected Poems,* ed. Gerald Hammond (Manchester: Carcanet, 1980), p. 51, line 459. Hereafter "Skelton."

"rhynchokinetic"—J. del Hoyo, A. Elliott, and J. Sargatal, eds., *Handbook of the Birds of the World,* vol. 3 (Barcelona: Lynx Edicions, 1996), p. 466. Hereafter *"HBW."*

"the writhings of a worm"—Yarrell, vol. 3, p. 329.

a diet of worms, eating almost its own body weight of them—*BWP,* vol. 3, p. 449.

I browse the entry for woodcock—Woodcock recipes from *Larousse Gastronomique* (1984; English trans., London: Paul Hamlyn, 1997), pp. 1160–61.

In September 1465 a banquet—Gurney, p. 87.

"Mr. Richardson of Bramshot"—Walter Johnson, ed., *Journals of Gilbert White* (1931; repr. Cambridge, MA: The MIT Press, 1971), pp. 143–44. Hereafter "White, *Journals.*"

"We were shooting above a railway cutting"—Lord Home, *Border Reflections* (London: Collins, 1979), p. 62.

"In the year 1833, a Woodcock"—Yarrell, vol. 3, p. 334.

the celebrated passage in "The Prelude"—William Wordsworth, *The Prelude,* 1805 text, ed. Ernest de Selincourt and revised by Stephen Gill (Oxford: Oxford University Press, 1978), book 1, pp. 9–10, lines 313–24.

Gilbert White reports—White, *Journals,* pp. 269–70.

"one was actually impaled on the weathercock"—Yarrell, vol. 3, p. 325.

"calling and apparently struggling to prop young between feet"—*BWP,* vol. 3, p. 453.

One is a staunch believer in the woodcock as porter—Home, p. 58, and Yarrell, vol. 3, pp. 326–28.

"edge of the orison"—Eric Robinson and David Powell, eds., *John Clare by Himself* (Manchester: Carcanet, 2002), p. 40.

knot coming and going—*HBW,* vol. 3, p. 477.

In 1212 in Cambridgeshire, on the marsh edge of the Wash, King John—Gurney, p. 49.

AUGUST: IN THE HAND

"We were all day hunting the wren"—Quoted in Edward A. Armstrong, *The Wren* (London: Collins, 1955), p. 5.

"now he must hold his hand"—From "St. Kevin and the Blackbird," *The Sprit Level* (London: Faber, 1996).

tied *"light silver threads" around the legs of the pewees*—John James Audubon, "The Pewee Flycatcher," in *The Audubon Reader,* ed. Richard Rhodes (London: Everyman's Library, 2006), p. 8. Hereafter *"Audubon Reader."*

"the sweet fruits of poetical despondence"—Jane Austen, *Persuasion,* ed. James Kinsley (1818; repr. Oxford: Oxford University Press, 2004), p. 72.

SEPTEMBER: LEAVING HOME

"Altogether elsewhere, vast"—The last stanza of W. H. Auden's "The Fall of Rome," in *Collected Poems,* ed. Edward Mendelson (London: Faber and Faber, 2007), revised edition, p. 331. Hereafter "Auden." The poem is dated January 1947.

"Tracing a memory"—W. S. Merwin, "Shore Birds," in *Selected Poems* (London: Bloodaxe, Tarset, 2007), p. 159. The poem was first published in 1999.

The storm or *gyndagooster*—Quoted by Hugh MacDiarmid in *The Islands of Scotland* (London: Batsford, 1939), p. 89.

"a hollow place in the rock like a Coffin"—From Coleridge's notebooks, quoted in Richard Holmes, *Coleridge: Early Visions* (London: Hodder and Stoughton, 1989), p. 304.

wind thrush—so called in Christopher Merrett's list of British birds, published in 1666, Peter Bircham, *A History of Ornithology,* (London: Collins, 2007), p. 53. Hereafter Bircham.

"a favourite with me"—William Eagle Clarke, *Studies in Bird Migration* (London and Edinburgh: Gurney and Jackson & Oliver and Boyd, 1912), vol. 2, p. 133.

"as it rose from some rough grass"—Ibid., p. 136.

"Poseidon sat at his desk"—Franz Kafka, "Poseidon," in *Descriptions of a Struggle and Other Stories*, trans. Malcolm Pasley (London: Penguin, 1982), p. 118.

OCTOBER: CAGE

"Hope Joy, Youth"—Charles Dickens, *Bleak House* (Oxford: Oxford University Press, 1996), p. 217.

No, no, no, no!—Shakespeare, *King Lear*, V.iii.9–10.

Aristotle thought that the summer redstarts—See John Buxton, *The Redstart* (London: Collins, 1950), p. 112. Hereafter "Buxton."

"In the summer of 1940"—Ibid., p. 1.

he *"owned the best handwriting"*—Peter Marren, *The New Naturalists*, 2nd ed. (London: Collins, 2005), p. 283.

"These redstarts . . . I loved"—Buxton, p. 4.

"I must be understood"—Ibid., pp. 3–4.

"I wish that some naturalist"—Ibid., p. 107.

"There is scarcely any point"—Ibid., p. 140.

The papers include—Some of John Buxton's notebooks and papers are held in the Alexander Library at the Edward Grey Institute in Oxford. The letters are in a box file labeled "Buxton 5." The letter from Huxley to Buxton is dated January 13, 1948, the one from Fisher to Buxton June 11, 1948.

"since it is by this that they recognize"—Buxton, p. 112.

"The action is performed so often"—Ibid., pp. 72–73.

"The displaying male"—Ibid., p. 30.

"Perhaps it would have been"—Ibid., p. 140

Fisher approves of John Clare—James Fisher, *The Shell Bird Book* (London: Ebury Press and Michael Joseph, 1966), p. 193.

the Molesworth books, a popular 1950s series—Geoffrey Willans and Ronald Searle, *Back in the Jug Agane* (1959; repr. Penguin, 2000), p. 313.

"Nor would I change"—Buxton, p. 140.

"Yet even now"—Ibid., p. 139.

a man of negative capability in Keats's term—John Keats, letter to George and Tom Keats, December 21–27, 1817, in *Letters: A Selection*, ed. Robert Gittings (Oxford: Oxford University Press, 1977), p. 43. Hereafter "Keats, *Letters.*"

"the way in which scientific people and their followers"—Edward Thomas, letter to George Bottomley, quoted by Edna Longley in notes to Thomas's poem "The Unknown Bird" in Edward Thomas, *The Annotated Collected Poems*, ed. Edna Longley (London: Bloodaxe, Tarset, 2008), p. 183. Hereafter "Thomas: *Annotated.*"

"Artis has one in his collection"—John Clare, Natural History Letter III, quoted in *John Clare's Birds*, eds. Eric Robinson and Richard Fitter (Oxford: Oxford University Press, 1982), p. 14. Hereafter *"Clare's Birds."*

He clasps the crag—Alfred Tennyson, "The Eagle," Christopher Ricks, ed., *The New Oxford Book of Victorian Verse* (Oxford: Oxford University Press, 1990), p. 19.

John Clare's poem on a yellowhammer's nest—"The Yellowhammer," *John Clare*, The Oxford Authors, eds. Eric Robinson and David Powell (Oxford: Oxford University Press, 1984), p. 417. Hereafter "Clare."

the description in The Handbook of British Birds—H. F. Witherby, F. C. R. Jourdain, N. F. Ticehurst, and B. W. Tucker, *The Handbook of British Birds* (London: H. F. & G. Witherby, 1943–44), second impression, vol. 1, p. 114.

Clare's note on the yellowhammer's eggs—*Clare's Birds*, p. 14.

out hunting—Walter Scott described in "Finding," Guy Davenport, *The Geography of the Imagination: Forty Essays*, (London: Picador, 1984), p. 365.

A single letter from D. H. Lawrence—D. H. Lawrence, letter to Lady Cynthia Asquith from Garsington Manor, Oxford, November 9, 1915, in D. H. Lawrence, *The Collected Letters*, ed. Harry T. Moore (London: Heinemann, 1977), vol. 1, p. 378.

Bruno Schultz's stories—See "Birds," a story in the collection *The Street of Crocodiles,* in *The Collected Works of Bruno Schultz,* ed. Jerzy Ficowski, trans. Celina Wieniewska (London: Picador, 1998).

Ted Hughes's crow poems—Ted Hughes, *The Life and Songs of the Crow* (London: Faber, 1974). See also Leonard Baskin's accompanying artwork of testicular corvids.

Joseph Cornell's box constructions—See Deborah Soloman, *Utopia Parkway: The Life and Work of Joseph Cornell* (Boston: MFA, 1997).

Peter Szöke, on his bizarre record—*The Unknown Music of Birds/Az Ismeretlen Madárzene* (Hungaroton LPX 19347, 1987).

Ian Dury—"Billericay Dickie" is the final track on side one of *New Boots and Panties!!,* released in 1977.

Rimbaud kept a list of pigeon names—See Graham Robb, *Rimbaud* (London: Picador, 2000), pp. 258–59.

Some of his Pisan cantos—Ezra Pound, *The Pisan Cantos,* ed. Richard Sieburth (New York: New Directions, 2003).

experiments on swallows—John J. Videler, *Avian Flight* (Oxford: Oxford University Press, 2005), p. 182.

Lockley reports—R. M. Lockley, *I Know an Island* (London: George G. Harrap, 1938), pp. 172 and 176.

The German word for migratory nocturnal restlessness—Ian Newton, *The Migration Ecology of Birds* (London: Academic Press, 2008), p. 338. Hereafter "Newton."

"inkpads were placed on the floor of their cages"—Ibid., p. 37.

"M Dr Hlln Grdnr"—John Clare, *Selected Letters,* ed. Mark Storey (Oxford: Oxford University Press, 1990), p. 220.

Clare had recorded how his son found a nightingale's nest—Clare's *Birds,* p. 45.

NOVEMBER: THE GORGE

"Look through my eyes up"—W. S. Graham, "Enter a Cloud," in *New Collected Poems,* ed. Matthew Francis (London: Faber, 2004), p. 216.

"Only one ship is seeking us"—Larkin, p. 294.
"At eleven o'clock"—J. A. Baker, *The Peregrine* (1967; repr. New York: New York Review Books, 2005) pp. 106–7. Hereafter "Baker."
"but it is too far"—Ibid., p. 120.
an aside on the crow catchers of Königsberg—Ibid., p. 27.
"tramps and gypsies"—Ibid., p. 28.
"I have found an unusual concentration"—Ibid., p. 28.
"inhibited by a code of behaviour"—Ibid., p. 26.
"The odd are always singled out"—Ibid., p. 32.
"I found myself crouching over the kill"—Ibid., p. 95.
"a man descending through the trap-door of a loft"—Ibid., p. 112.
"Imprisoned by horizons"—Ibid., p. 170.
described the wood warbler as "extremely abundant"—Herbert C. Playne, *Some Common Birds of the Neighbourhood of Clifton* (Clifton and London: J. Baker & Simpkin Marshall, 1907). The redstart is described as "fairly common in the oak trees of Nightingale Valley" of Leigh Woods. I have never seen one there.

DECEMBER: BLACK BIRDS AND BLACK NIGHTS

"An Airy Crowd came rushing where he stood"—*Virgil's Aeneid*, trans. John Dryden, ed. Frederick M. Keener (1697; repr. London: Penguin, 1997), book 6, lines 422–32, p. 159.
"as we take, in fact, a general view"—William James, *The Principles of Psychology*, vol. 1 (New York: Dover, 1950), pp. 243–44.
the Sweet Track—Oliver Rackham, *Trees and Woodland in the British Landscape*, 1st ed. (London: J. M. Dent, 1983), p. 48, and revised ed. (London: Phoenix, 1996), p. 37.
Eight million were counted—David K. Ballance, *A History of the Birds of Somerset* (Penryn: Isabelline Books, 2006), p. 310.
"Myre dromble"—Gurney, p. 183: "literally it means the sluggish bird of the marshes, but corrupted as it soon became into the shorter name of miredrum, it signifies the bird which drums or booms."
"Hi! Shoo all o' the birds"—James Reeves, ed., *The Idiom of the People* (London: Mercury House, 1962), p. 76. Hereafter "Idiom."

Nero's starling—Pliny the Elder, *Natural History: A Selection*, trans. John F. Healy (London: Penguin, 2004), p. 147. Pliny also reports a Roman raven, the pet of a shoemaker, that could talk too but was killed by another cobbler for shitting on his shoe; it was then given a funeral of great pomp, "the draped bier carried on the shoulders of two Ethiopians" (p. 148).

Bobby Tulloch's scientific note—See Ian Wallace, *Birdwatching in the Seventies* (London: Macmillan, 1981), p. 34, and Bobby Tulloch, *Migrations: Travels of a Naturalist* (London: Kyle Cathie, 1991), p. 3.

Rosa Luxemburg—See John Berger, *Here Is Where We Meet* (London: Bloomsbury, 2005), p. 227: "Rosa loved birds—particularly the urban starlings who fly en masse above the streets and over the roofs."

Starlings in Dante—See Dante, *The Inferno of Dante Alighieri*, trans. Ciaran Carson (London: Granta, 2002), p. 32, lines 40–43.

Edward Thomas in France—Edward Thomas, "War Diary" in Thomas, *Collected Poems*, p. 165.

toward Central Park, where they were first released—Christopher Lever, *Naturalised Birds of the World* (London: T. & A. D. Poyser, 2005), p. 195.

Rachel Carson standing up—See *Lost Woods: The Discovered Writing of Rachel Carson*, ed. Linda Lear (Boston: Beacon Press, 1998), p. 24.

An English roost of starlings—See Peter Bircham, *A History of Ornithology* (London: Collins, 2007), p. 308. Hereafter "Bircham."

"abundant excrement"—Gurney, p. 232.

Laurence Sterne's caged starling—Laurence Sterne, *A Sentimental Journey and other Writings*, ed. Tom Keymer (London: J. M. Dent, 1994), p. 60.

"Starlings!" shouted derisively—Thomson, p. 139.

Coleridge keeping starlings in mind—Letter to William Godwin, 1802, in *Samuel Taylor Coleridge: Selected Letters*, ed. H. J. Jackson (Oxford: Oxford University Press, 1988), p. 101.

Coleridge, two years before that—*Coleridge's Notebooks: A Selection*, ed. Seamus Perry (Oxford: Oxford University Press, 2002), p. 39.

endorsing summer—Louis MacNeice, "Nature Notes—Corncrakes," MacNeice, p. 549.

sourdine in their throat—Andrew Marvell, "Upon Appleton House," Marvell, p. 66.

The "brabbling" starling—Skelton, p. 51, line 461.

W. B. Yeats's "stare"—"The Stare's Nest by My Window," poem VI of "Meditations in Time of Civil War," in *Collected Poems* (1928; repr. London: Macmillan, 1978), p. 230.

John Clare's "starnals"—"Autumn evening," Clare, p. 241.

"a green thought in a green shade"—Marvell, p. 48.

Bede and his image of life's journey as a sparrow—Bede's sparrow is in his *Ecclesiastical History of the English People*. Wordsworth adapts the idea in his "Ecclesiastical Sonnet XVI": "Man's life is like a Sparrow . . . / Fluttering, / Here did it enter; there, on hasty wing, / Flies out . . ." William Wordsworth, *The Poems*, vol. 2, ed. John O. Hayden (London: Penguin, 1977), p. 453.

"If a Sparrow come before my window"—John Keats, letter to Benjamin Bailey, November 22, 1817, in Keats, *Letters*, p. 38.

JANUARY: SOUTH

"You acid-blue metallic bird"—From "The Blue Jay," in *The Complete Poems*, eds. Vivian de Solo Pinot and Warren Roberts (Harmondsworth: Penguin, 1980), p. 375.

"What birds were they?"—James Joyce, *A Portrait of the Artist as a Young Man*, ed. Seamus Deane (1914–15; repr. London: Penguin, 1992), p. 245.

"the three orange trees"—Antoine de Saint-Exupéry, *Wind, Sand and Stars*, trans. William Rees (1939; repr. London: Penguin, 2000), p. 8.

It takes a willow warbler between twenty-nine and forty-four hours— Newton, pp. 149–50

of 290,000 ringed in Britain and Ireland—See Chris Wernham, Mike Toms, John Marchant, Jacquie Clarke, Gavin Siriwardena, and Stephen Baillie, eds., *The Migration Atlas* (London: T. & A. D. Poyser, 2002), pp. 465–67.

"pendent bed and procreant cradle"—Shakespeare, *Macbeth*, I. vi. 8

"Ready-made nests"—John Clare, "Childhood," in *Selected Poems*, ed. Jonathan Bate (London: Faber, 2004), p. 108.

"birds like to nestle"—"Autumn Birds," Clare, p.267.

"Tis Spring"—"Birds Nests," Clare, p. 427.

"Sir John Lawrence, Kt"—Daniel Defoe, *Journal of the Plague Year* (1722; repr. London: Penguin, 1966), p. 118. Hereafter "Defoe."

"the driver being dead . . . and the horses running too near it"—Ibid., p. 192.

"great numbers went out of the world"—Ibid., p. 116.

the greater honeyguide has even been recorded *"eating candles"*—Frank B. Gill, *Ornithology*, 3rd ed (New York: W. H. Freeman, 2007), p. 171. Hereafter "Gill."

"looks on at the humans"—*HBW*, vol. 7, p. 279.

human honey-guiding *"is not a genetically deeply embedded"*—Ibid., p. 280.

FEBRUARY: CRONK

"Why is a raven like a writing desk?"—Lewis Carroll, *Alice's Adventures in Wonderland and Through the Looking Glass* (London: Penguin, 1998), p. 60.

"Hoarse with fulfilment"—W. S. Merwin, "Noah's Raven," in *Migration: New and Selected Poems* (Port Townsend: Copper Canyon Press, 2005), p. 84.

sixty-four have been recorded—Derek Ratcliffe, *The Raven* (London: T. & A. D. Poyser, 1997), p. 265. "The oft-repeated statement that the species utters no less than 64 recognisably different calls appears to date from the days of the Roman augurs, to whom they had immense significance."

"We shouldn't forget the Goofus Bird"—Jorge Luis Borges, *The Book of Imaginary Beings*, trans. Norman Thomas di Giovanni (London: Penguin, 1974), p. 69.

"Like dominoes"—Baker, p. 88.

"The raven himself"—Shakespeare, *Macbeth*, I.v.38

"Ye'll sit on his white hause-bane"—"The Twa Corbies," in *The Oxford Book of English Verse,* ed. Christopher Ricks (Oxford: Oxford University Press, 1999), p. 295.

A newspaper in January 1767 describes a blacksmith—Clifford Morsley, *News from the English Countryside* (London: Harrap, 1979) p. 58.

Molly Bloom was from there and remembered its "Moorish wall"—See James Joyce, *Ulysses* (London: Penguin, 1992), pp. 932–33). Her memories of the rock come right at the end of her monologue.

Gilbert White knew—See White's letter XXXIII, 1770, in *The Natural History of Selborne,* ed. James Fisher (London: The Cresset Press, 1966), pp. 75–76. Hereafter "White."

Konrad Lorenz wrote about his pet ravens, especially Roah—Konrad Lorenz, from an article in *The Countryman,* reprinted in *The Bird-Lover's Bedside Book,* ed. R. M. Lockley (London: The Country Book Club, 1960), p. 26.

Arthur Streeb-Greebling—aka Peter Cook, "We're knee-deep in feathers off that part of the coast . . . not a single success in forty years of training." The internet widely publishes these scripts.

MARCH: FEATHERS AND BONES

"The bird is dead"—Shakespeare, *Cymbeline,* IV.ii.197–98

"It was a fine day and K. felt like going for a walk"—Franz Kafka, "A Dream," from the story collection *A Country Doctor,* reprinted in *The Transformation and Other Stories,* trans. Malcolm Pasley (London: Penguin, 1992), p. 185.

"Birds too have died in great numbers" —Roger Durman, ed., *Bird Observatories in Britain and Ireland* (Berkhamsted: T. & A. D. Poyser, 1976), p. 38.

Eagle Medicine—Man, the Crow Indian, photographed by Edward S. Curtis—See *The North American Indian: The Complete Portfolios* (Köln: Taschen, 1997), p. 185.

Edward Thomas's poem "Birds' Nests"—Thomas: *Annotated,* p. 43.

Ornithologia—See Bircham, p. 98.

"I had no paper"—Buxton, p. 1.

Allan Octavian Hume—See Michael Walters, *A Concise History of Ornithology* (London: Christopher Helm, 2003), p. 157. The photograph of Hill on the same page shows a balding man in a wing collar with bushy gray eyebrows and a dropping bird's-nest mustache. It is hard to read the catastrophe anywhere but in his half-closed eyes.

"A Collection of Birds' Feet"—Bruce Taggart ("YOC member"), *Bird Life* 6, no. 4 (October-December 1970), p. 117.

"Make the boy interested in natural history"—Robert Falcon Scott, *Journals*, ed. Max Jones (Oxford: Oxford University Press, 2005), p. 419.

"being in the garden at Bockhampton"—Florence Hardy, *The Life of Thomas Hardy* (London: Studio Editions, 1994), p. 263.

APRIL: A SINGING WORLD

"Aria Spontanea . . . An Air that whizzed"—Samuel Taylor Coleridge, "Aria Spontanea," in *Selected Poems*, ed. Richard Holmes (London: HarperCollins, 1996), p. 257. Hereafter "Coleridge." The odd word, "mute," means bird shit. It comes from the French *émeut* and is first recorded in English in 1575 (Shorter Oxford English Dictionary).

New song patterns in dreams—Gill, p. 207.

Keats did something to the nightingale in his ode to the bird—See Keats, pp. 285–88.

John Clare's description of nightingale song in "The Progress of Rhyme"—Clare, pp. 158–59. Lines in the poem are remarkably accurate field notes or transcriptions: "Wew-wew wew-wew chur-chur chur-chur / Woo-it woo-it—could this be her / Tee-rew Tee-rew tee-rew—tee-rew / Chew-rit chew-rit—and ever new / Will-will will-will grig-grig grig-grig."

poem by Ivor Gurney called "The Nightingales"—In Ivor Gurney, *Best Poems and the Book of Five Makings*, eds. R.K.R. Thornton and George Walter (Ashington and Manchester: Mid Northumberland Arts Group and Carcanet, 1995), pp. 135–36.

Nabokov dealing somehow with nightingales—Vladimir Nabokov, *Speak, Memory* (London: Penguin, 2000), p. 67.

MAY: INTO AND OUT OF THE HOLE

"There was a tree"—Sung by Mrs. Grace Coles at Enmore, West Somerset, 1906, collected in Cecil Sharp, *Idiom*, p. 211. The village of Enmore is the site of the first free school in England, founded by John Poole in 1810; perhaps Grace Coles studied there. Poole was the nephew of Thomas Poole of Nether Stowey and a friend of Coleridge and Wordsworth. They might have known the song.

"But enough of greenwood's gossip"—James Joyce, *Finnegans Wake* (London: Penguin, 1992), p. 450.

"everywhere the blue sky belongs to them"—S. T. Coleridge, gloss to "The Rime of the Ancient Mariner," part IV, line 263, in Coleridge, p. 89.

"from the mouth of a swift"—The Rev. F. O. Morris, *British Birds: A Selection from the Original Work,* ed. Tony Soper (London: Spring Books, 1987), p. 78.

"As if the bow had flown off with the arrow"—"Haymaking," Thomas: *Annotated*, p. 95.

"The Sea and the Mirror"—Auden, p. 405.

"the verb startle"—See Francesca Greenoak, *All the Birds of the Air: The Names, Lore and Literature of British Birds* (Harmondsworth: Penguin, 1981), p. 239.

"The Firetails Nest"—Clare, p. 212.

Edward Thomas's "Adlestrop," Thomas: *Annotated*, p. 51.

Older blackbirds sing in the evening more than in the morning—David Snow, *A Study of Blackbirds,* rev. ed. (1958; repr. London: British Museum [Natural History], 1988), pp. 191–92.

"scurr"—See Joanna Cullen Brown, *Let Me Enjoy the Earth: Thomas Hardy and Nature* (London: Allison and Busby, 1990), p. 100.

"croo"—*Clare's Birds*, p. 64.

Gilbert White feels a nightjar singing—Gilbert White, letter XXII, 1769, in White, p. 55.

Its coloring, if we could ever see it, is extraordinary—*BWP*, vol. 4, pp. 633–34.

the nightjar was lodged at the bottom of plate 46—Roger Peterson, Guy Mountfort, and P.A.D. Hollam, *A Field Guide to the Birds of Britain and Europe,* 11th ed. (London: Collins, 1967), plate 46, p. 181.

Gilbert White was given a clutch of two nightjar eggs—White, *Journals,* p. 338.

In August 1786 Gilbert White had a young nightjar—Ibid., pp. 281–82.

living lamps—Andrew Marvell, "The Mower to the Glow-worms," Marvell, p. 43.

AFTERWORD: SINGING

"The wild duck startles like a sudden thought"—John Clare, "Autumn Birds," Clare, p. 267.

"When an individual [passenger pigeon] is seen gliding through the woods"—*Audubon Reader,* p. 65.

"A palm at the end of the mind"—"Of Mere Being," in *Opus Posthumous* (London: Faber, 1959), pp. 117–18.

PERMISSIONS

"The Fall of Rome," copyright 1947 by W. H. Auden, "The Sea and the Mirror," copyright © 1976 by Edward Mendelson, William Meredith and Monroe K. Spears, Executors of the Estate of W. H. Auden., from COLLECTED POEMS by W. H. Auden. Used by permission of Random House, Inc.

John Buxton, excerpts from *The Redstart.* Copyright 1950 by John Buxton. Reprinted with the permission of HarperCollins, Ltd.

W. S. Graham, excerpt from "Enter a Cloud" from *New Collected Poems,* edited by Matthew Francis. Reprinted with the permission of Michael and Margaret Snow, Literary Executors for the W. S. Graham Estate.

Ivor Gurney, excerpts from "The Nightingales" from *Best Poems and the Book of Five Makings,* edited by R.K.R. Thornton and George Walter. Copyright © 1995 by the Ivor Gurney Estate. Reprinted with the permission of the Carcanet Press, Ltd.

D. H. Lawrence, excerpt from "The Blue Jay" from *The Complete*

Acknowledgments

My deepest thanks are for my father. I couldn't have had a more fortunate beginning as a bird-watcher. My parents gave me my first pair of binoculars when I was seven, and my dad took me bird-watching countless times through my childhood, long after his own interest must have been exhausted. His patience, generosity, and enthusiasm were matchless. He bought me most of the books I have cited here, often many years before I realized their value, and his tolerance of my requests—taking me and my earnest bird-watching friends out on day trips in atrocious weather, covering my paper route for me for months so I could chase rarities, and driving me across the country many times looking for avocets and nightingales, a black vulture, and a scops owl—made him a perfect dad for me. That is still the case. My mother was lovingly tolerant of my obsession and my calls upon my father. My sister put up with regular and savage attacks—rabbit punches and "Chinese burns"—on her back and arms when she failed my impromptu bird identification tests. I am sorry, Jenny, though I am sure you are better now on your finches than you would have been.

My own children have been more than kind. A bird-watching dad is dire, but they have sweetly accepted my absences and my erratic driving (frequent stops, reversals, and muddy tracks) and bizarre holiday plans (hot Spanish plains, Scottish islands without toilets, bird-watching in Los Angeles). Dominic's comically willful ornithological ignorance is a regular and happy corrective to my enthusiastic excesses. Lucian is, whether he wants to be one or not, a born naturalist and has been great company on many trips, as well as an ace finder of dead things and grabber of the living. Their mother, Stephanie Parker, has been incredibly accommodating to the demands of this book, and much else besides; the words here began in many places—one such was a walk with her across Exmoor in the early summer more than fifteen years ago, when redstarts, wood warblers, and pied flycatchers retook possession of my imagination. In the middle of her (briefly) difficult pregnancy, we walked the Doone Valley as if we were both carrying a glass of water filled to its brim. I won't forget it. She also introduced me to Thomas Tallis. I won't forget that.

Many others have helped me. My writing began thanks to Marybeth Hamilton, who asked me to contribute to the *History Workshop Journal*, and Redmond O'Hanlon, who did the same for the TLS. Lavinia Greenlaw and Claire Armitstead helped me on my way in the UK and Eliot Weinberger and Bradford Morrow did the same in the USA. Andrew McNeillie of *Archipelago* magazine has kept me going as I have been writing. Some of the words here first appeared in various forms in the *HWJ, Conjunctions,* and *Archipelago*. I am grateful to all at those journals.

Mark Cocker, as well as being a superb birder with the sharpest ears around and a writer of great bird words, is also brilliant company for dirt tracks in Morocco, dawn treks in Turkey, and drenchings on Rum. He is the best possible bird friend. Greg Poole keeps me on my bird toes in Bristol, as well as being the funniest man to go bird-watching with and the finest bird artist of our time. Adam Nicolson is the kindest of men, as well as wonderfully stimulating company and the owner of the most marvelously abandoned laugh. A long time ago, through the autumns of the 1970s, Tom Nichols and Antony Merritt let me accompany them on their endless circumnavigations of Chew Valley Lake as they counted ruffs. In part this book is for them—in those days I couldn't have explained it but much was being laid down. Tom is a wonderful cousin too, and though our designs, also from the 1970s, for the animal collecting vehicle to be called the Deenick 2000 never left the drawing board ("packaway cages that packaway," etc.), I hope he finds some of those dreams in these pages. I also thank Nigel Collar and Martin Jenkins, two onetime colleagues from my short-lived career in the early 1980s as a professional conservationist. Both are more literate and literary than me but both, in their different deep antipathies to nature writing, kept me from starting this book. Since I couldn't have done it before, I thank them both for the drunken arguments that kept me going through the lost years. Paul Dodgson and Ken Arnold have made fun of me and my birds too in the friendliest way imaginable, in Ken's case for nearly thirty years; I love them for it.

At the BBC a friendly and supportive team of colleagues has made being a radio producer who wants to write a reality. My particular thanks go to Clare McGinn, Kate Chaney, and Ali Serle, and to Iain Hunter and Mike Burgess—"let money and work be as casual in human life as they are in a bird's life—damn it all" was how D. H. Lawrence put it.

Specific thanks go to the following: on the Wash, Peter and Anne Welberry Smith and the crew of the Eastern Sea Fisheries *Protector III* patrol boat; at Chew Valley Ringing Group, my trainer Robin Prytherch and the other ringers; on Fair Isle, the skipper of the *Good Shepherd,* Neil Thompson, the observatory wardens Iain Robertson and Deryk Shaw and their expert teams, and Mark Ward and Adrian Cooper, who enlivened my second stay; in Zambia, Ian Bruce Miller and Emma Bruce Miller, Lazaro Hamusikili, Kiverness Moto and Collins Moya, Ailsa Green and the late Major John Colebrook-Robjent and his wife Royce (the major died as the book was going to press—it was a privilege to have met him); at the Bristol and Gloucestershire Gliding Club, Tim Allen; in the Brecks, Paul Dolman; Helen Macdonald shared her work on John Buxton and her eclectic brilliance on all bird things.

Acknowledgments

Much of the bird-watching in the book was done on my own. Some was done in company; all of it was informed by talking to others. For multiple stimulations, on the page and on location, I thank Fraser Harrison, Richard Holmes, Zinovy Zinik, Robert Macfarlane, Harry Cory Wright, and Richard Mabey. Simon Armitage and Paul Farley both came out as birders in the time that I have known them with characteristic and inspirational brilliance and much comedy. Michael Longley and Peter Reading already were great bird poets and remain lodestars. More than anyone, Kathleen Jamie challenged me brilliantly by writing the most original poetry on how we all live in nature as well as being excellent company on various saturated Scottish islands and the tossing boats between them. That the names on this list are not mentioned within my chapters only means I am terrified that all I say has been said before and better by them.

Some people read the entire book or part of it as it came in to land. For their readings, suggestions, and encouragements I am enormously grateful to Susannah Clapp, Tessa Hadley, Jeremy Harding, Andrew Motion, and Robin Robertson. To have such writers as friends and friendly editors has made me very happy as well as improving my prose beyond measure.

Pat Kavanagh, my agent until her death, was my first reader, and without her encouragement and acumen the book wouldn't exist. It is awful and sad that she didn't see the end of what she had started and read only part of the finished work. I will treasure, though, her enthusiasm for what she had read and a shared evening of hopeless bird-watching on Hampstead Heath with shrieking parakeets and screaming swifts and almost nothing else until some exquisite Banyuls rosé wine came into view. Sarah Ballard, who worked alongside Pat, has taken the book on with wonderful enthusiasm and was exceptionally helpful. Zoë Pagnamenta has guided me brilliantly in this regard into the New World.

Much of this book was assembled and rewritten in a beautiful room at the top of a stone tower at the Santa Maddalena Foundation in the Tuscan countryside. Wild boar hunters in the woods kept me indoors and every morning a green woodpecker stapled my head into place, furiously tapping around the eaves outside. Without them, the instant friendship there of Nayla El Amin and the amazing trusting openness of Beatrice Monti della Corte I would have foundered.

The teams at the Free Press and Jonathan Cape have been extremely good to me. I thank Tim Waller and Donna Loffredo. I met my American editor, Leslie Meredith, for the first time when we had lunch in a smart Manhattan restaurant. I knew everything was going to go well when we skipped coffee, she pulled out her binoculars from her handbag, and we headed out into Central Park. She has been enormously helpful and encouraging. Dan Franklin, in the UK, manages to be both everything you want and nothing that you expect in an editor. He has a unique way of being exacting by being generous at the same time, so that getting the words right for him becomes the most important thing. I hope I have.

I didn't think it was possible to fall in love with a bird-watcher, and I hadn't when I started this book. I began it writing about the impossibility of a shared life. Claire Spottiswoode has changed my mind, my heart, my bird list, and my world.

About the Author

Tim Dee is a BBC radio producer and writer. Born in Liverpool, he studied in Cambridge and Budapest and worked for the International Council for Bird Preservation before joining the BBC, where he makes features and documentaries and directs plays. He is currently editing *The Poetry of Birds* with Simon Armitage. He has been watching birds since he was three.